How animals work

by

Knut Schmidt-Nielsen

James B. Duke Professor of Physiology in the Department of Zoology, Duke University

Cambridge
at the University Press

Published by the Syndics of the Cambridge University Press
Bentley House, 200 Euston Road, London NW1 2DB
American Branch: 32 East 57th Street, New York, N.Y.10022

© Cambridge University Press 1972

Library of Congress Catalogue Card Number: 77-174262

ISBNS:
0 521 08417 2 clothbound
0 521 09692 8 paperback

Typeset in Great Britain by
William Clowes & Sons Limited
London, Colchester and Beccles

Printed in USA

First published 1972
Reprinted 1973

Contents

Foreword

This book is aimed at undergraduate students of vertebrate physiology and biology but, as is the case with previous books written by Professor Knut Schmidt-Nielsen, professional biologists of all age groups are likely to change their outlook and views in the course of reading it, sometimes substantially. Why is this? First, because style and contents make up such an integrated whole that I for one felt compelled to read the book from beginning to end in one go. The major reason, however, is that, although firmly rooted in classical physiology and morphology, Schmidt-Nielsen is never conventional and is able to see new and surprising solutions and relationships. It is the mixture of classical scholarship and common sense with the innovating spirit of an artist which has led him and his collaborators to an understanding of how animals survive in deserts, the existence and function of the salt glands in birds and reptiles and, as it now seems, to the solution of a classical problem, the function of the bird's lung.

For the zoologist as well as for the general reader, the book has another charm. It is written by a true cosmopolitan who has brought together personal experience and examples from five continents, and, in spite of all this, one can almost hear through the lines the characteristic Norwegian accent. Born in Trondheim in 1915, Knut Schmidt-Nielsen went to school in Norway, studied at the University of Copenhagen under August Krogh and did the work for his D.Phil. at the Carlsberg Laboratories under Kaj Linderstrøm-Lang. But it was his first encounter with the Arizona Desert in 1946 which set the scene for most of his later research activities. Since 1952 he has been Professor of Physiology in the Department of Zoology at Duke University. The Managers could hardly have made a more suitable choice than to appoint him the Hans Gadow Lecturer at the University of Cambridge for 1970–1 and so provide a formal framework for the lectures on which this book is based, as well as a strengthening of the ties of friendship between the two universities.

TORKEL WEIS-FOGH
Professor and Head of
Department of Zoology,
University of Cambridge, England

To

STUDENTS
FRIENDS
COLLEAGUES

In preparing the Hans Gadow Lectures I have drawn on work of many students, friends, and colleagues with whom I have had the privilege to be associated, and their contributions weigh heavily in the list of references. Nevertheless, I also wish to express here the immense debt of gratitude I owe to my collaborators, W. L. Bretz, W. A. Calder, Jr, F. G. Carey, J. C. Collins, J. E. Cohn, E. C. Crawford, Jr, R. Dmi'el, M. A. Fedak, B. Gjönnes, F. R. Hainsworth, D. S. Hinds, T. R. Houpt, D. C. Jackson, J. Kanwisher, J. L. Larimer, R. C. Lasiewski, P. L. Lutz, D. E. Murrish, P. Pennycuick, T. C. Pilkington, H. D. Prange, J. L. Raab, B. Schmidt-Nielsen, and C. R. Taylor. Many other collaborators, although not named here, have been equally important in the formulation of what I am trying to say in this book.

K. S. N.

1. Respiration and evaporation

Introduction

A simple biological problem may arouse our interest, but as we gain more knowledge the questions ramify and appear to grow in complexity. This may take us to new and seemingly unrelated problems, but in retrospect they are all related to the desire to find out how things work. If we are fortunate we will gain some insight, and when we understand underlying principles, the greatest reward seems to be in the simplicity of the answers.

Twenty-five years ago I had an opportunity to visit the Arizona deserts, and I was much impressed by their abundant animal life, which for me was quite unexpected. In particular, it surprised me that small rodents were quite numerous, in spite of the obvious absence of drinking water. What impressed me equally was that virtually no scientific information, other than a few anecdotal accounts, was available on how these animals could live seemingly without water. It has since become abundantly clear that the water balance of a desert rodent, such as a kangaroo rat, is quite like that of any other animal, that is, for the animal to remain in water balance, the intake of water must equal the total losses. It has become equally clear that they do not eke out a marginal existence during the long periods of drought; they manage as well as any animal that is at home in its natural environment.

Let us briefly consider the water balance of kangaroo rats, which live on seeds and other dry plant material. They do not usually drink, and they do not particularly seek out green and succulent plants to obtain water. A simplified account of their water balance will clarify the questions to be asked (table 1). Since these animals live on dry food, the amount of free water in the food is quite small, and the water which is derived from oxidation of the foodstuffs ('metabolic water') is nearly all that is available. This fairly small amount of water must suffice to cover all losses, which are by

Table 1. *Water balance of the kangaroo rat. For an animal to remain in water balance, the total gains must equal the total losses*

Gains		Losses	
Oxidation water	90%	Evaporation	70%
Free water in food	10%	Urine	25%
Drinking	0%	Feces	5%

evaporation, in the urine, and with the feces. The two last items can be minimized by excreting a highly concentrated urine and very dry droppings, but the means of reducing evaporation are more limited because the respiratory membranes are moist and air is exhaled saturated with water vapor. As a consequence, evaporation is the major and inescapable avenue for water loss in these animals. For a man in the desert evaporation is far greater because he uses sweating to remain cool in the hot surroundings. For desert rodents the available water would soon be exhausted if they were to sweat; actually, they have no sweat glands and avoid the need to sweat by staying in their relatively cool burrows during the hot desert day, restricting their active life to the cooler nights.

Evaporation, mice and men

We usually assume that air is exhaled saturated with water at close to body temperature, which for mammals and birds is some 37 °C. For reptiles the situation is different because they are 'cold-blooded', which means that their body temperature varies with that of the surroundings. Their temperature, and the water content of the exhaled air therefore, is much more variable, and usually also much lower than in birds and mammals.

When comparing the evaporation from different animals, it will be convenient to use some suitable common measure, for animals are of widely different sizes and their need for oxygen (and thus the ventilation of the lungs) varies greatly. Since lung ventilation, and thus the evaporation, varies with the call for oxygen, it is convenient to examine the amount of water evaporated relative to one unit volume of oxygen taken up. This becomes particularly useful when we consider that the amount of oxygen taken up is a direct measure of the amount of oxidation water formed in metabolism, which again stands for virtually the entire gain of water for the desert rodents we have under consideration.

If we now compare the evaporation of various animals (table 2) we find that those that normally live in deserts evaporate substantially less than non-desert animals or man. Why this difference?

Table 2. *Evaporation in rodents. To permit comparisons, the evaporation is expressed relative to the simultaneous oxygen consumption. For rodents, the figures represent total evaporation from lungs and skin combined, for man from the lungs only.* (Data from Schmidt-Nielsen, 1964)

Species	Evaporation mg H_2O ml^{-1} O_2 cons.
Kangaroo rat	0.54 ± 0.04
Pocket mouse	0.50 ± 0.07
Golden hamster	0.59 ± 0.05
White rat	0.94 ± 0.09
White mouse	0.85 ± 0.07
Man (lungs only)	0.84

Presumably both body temperature and lung ventilation are similar in the various species, so how do the differences arise? It should be noted that the figures in this table give the evaporation from the entire animal, and therefore include evaporation from the skin as well as from the lungs, except for man where skin evaporation is considerable and highly variable. It has been said that evaporation from the skin of rodents is small as they have no sweat glands, but in reality it is a substantial amount for white rats, in which nearly one-half of the evaporation is from the skin (Tennent, 1946). If this amount were subtracted, the remaining evaporation for the white rat would be similar to that in the other small rodents, thus eliminating the difference between desert and non-desert species.

The respiratory evaporation as given for man represents air saturated at close to body temperature. If we use this as our standard, the question becomes: How is it possible for other animals to reduce this evaporation? The possible solutions we can imagine would be; (1) that more than the usual amount of oxygen is extracted from the air in the lungs before it is exhaled, or (2) that the exhaled air is not saturated at body temperature.

The first possibility would seem quite reasonable, in fact more likely than the second. More oxygen could be removed in the lung if the oxygen affinity of the blood hemoglobin were increased. We know that in some animals this may be so; the llama, for example, has hemoglobin with a high oxygen affinity which enables it to take up oxygen more easily in the rarified atmosphere of high altitude in the Andes where it lives (Hall, Dill and Guzman Barron, 1936). A higher oxygen affinity is expressed as a shift to the left of the oxygen dissociation curve, but no such shift is found for kangaroo rat blood; on the contrary, the dissociation curve is located rather to the right, showing a low oxygen affinity as is usual for mammals of small body size.

This alone does not prove that there can be no increase in oxygen extraction in the lung, but other evidence strongly supports it. If, say, twice the usual amount of oxygen were to be removed from the lung air, the carbon dioxide concentration in the exhaled air would become twice the usual. Thus the blood carbon dioxide tension would be doubled, which would be reflected either as severe acidosis or as a change in the bicarbonate buffer system. The blood buffer system of kangaroo rats is as in other mammals, and we must therefore abandon the hypothesis of an increased oxygen extraction (Gjönnes and Schmidt-Nielsen, 1952).

We must now consider the second possibility, that the exhaled air is not saturated at body temperature. On examination we find that although the air is saturated, its temperature is far below body temperature. In the lung the air is at body-core temperature, on its passage from the lung to the outside it undergoes substantial cooling, and in the process water vapor is recondensed on the walls of the passageways.

A cold nose

If we measure the temperature inside the nasal passage of a kangaroo rat, we find the lowest temperature at the tip of the nose, with a longitudinal gradient extending inwards until body temperature is reached at a depth of about 20 mm (figure 1). When the animal breathes saturated air, the temperature at the tip of the nose approaches that of the surroundings; in dry air, however, the temperature at the tip is several degrees below the ambient air temperature. The air that flows through the nasal passages attains, within a few tenths of a degree, the same temperature as the wall, and an animal which inhales dry air thus exhales air at a temperature even lower than the inhaled. We should remember, of course, that the air as it leaves the lung is at body temperature, 38 °C, and is cooled as it flows out through the nose.

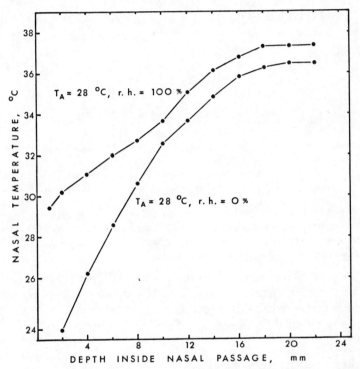

Figure 1. Temperatures in the nasal passage of the kangaroo rat (*Dipodomys merriami*). *Upper curve:* breathing air of 100% relative humidity at 28 °C. *Lower curve:* breathing completely dry air, also at 28 °C. Body temperature 38 °C. (From Jackson and Schmidt-Nielsen, 1964.)

To understand how this temperature gradient along the nasal passageway is established, it is useful to look at a simple model of the nose (figure 2). Let us assume that there is no evaporation to complicate things. During inhalation the walls of the passageway are cooled by the air passing by, and the wall temperature will approach that of the inhaled air, in this case, 28 °C. During exhalation, as warm air from the lungs passes over these cool surfaces, the air is cooled, and if the heat exchange is complete, the exhaled air temperature will approach that of the surface,

28 °C. The measurements on the kangaroo rat in moist air (see figure 1, upper curve) showed approximately this situation, the exhaled air temperature was nearly that of the ambient air. In completely dry air (also at 28 °C) the exhaled air was about 23 °C,

INHALATION   EXHALATION

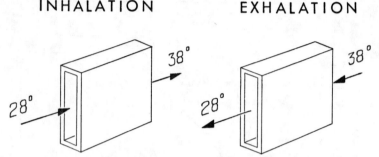

Figure 2. Model of heat exchange in the nasal passageways. Ambient air is 28 °C and saturated, body temperature is 38 °C. As inhaled air flows through the passageways (*left*) it gains heat and water vapor, and is saturated and at 38 °C before it reaches the lungs. On exhalation (*right*) the air flows over the cool walls, gives up heat, and water recondenses. As heat and water exchange approaches completion, exhaled air approaches 28 °C, saturated. (From Schmidt-Nielsen *et al.* 1970*b*.)

or 5 degrees below ambient air temperature (figure 1, lower curve). This is because during inhalation water evaporates from the moist nasal mucosa, which therefore becomes cooler than the air, just as the 'wet-bulb' thermometer of a sling psychrometer.

Flukes and flippers, or
Countercurrent heat exchange

The heat exchange in the nose has a great similarity to the well-known countercurrent heat exchange which takes place, for example, in the extremities of many aquatic animals, such as in the flippers of whales and the legs of wading birds. The body of a whale that swims in water near the freezing point is well insulated with blubber, but the thin streamlined flukes and flippers are uninsulated and highly vascularized and would have an excessive heat loss if it were not for the exchange of heat between arterial and venous blood in these structures. As the cold venous blood returns to the body from the flipper, the vessels run in close proximity to the arteries, in fact, they completely surround the artery, and heat from the arterial blood flows into the returning venous blood, which is thus re-heated before it enters the body (figure 3). Similarly, in the limbs of many animals both arteries and veins split up into a large number of parallel, intermingled vessels each with a diameter of about 1 mm or so, forming a discrete vascular bundle known as a rete (Scholander, 1958; Scholander and Krog, 1957). Whether the blood vessels form such a rete system, or in some other way run in close proximity, as in the flipper of the whale, is a question of design and does not alter the principle of

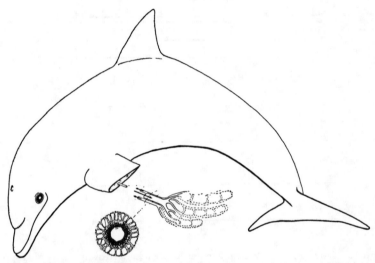

Figure 3. Heat loss from the extremities of whales is reduced by a special arrangement of the blood vessels. This cross-section of a porpoise flipper shows how an artery is completely surrounded by veins. As a result, the warm arterial blood gives up heat to the venous blood, which thus is pre-heated before it re-enters the body. (Drawing by D. E. Murrish, reproduced by permission from Schmidt-Nielsen, 1970.)

the heat recovery mechanism. The blood flows in opposite directions in arteries and veins, and heat exchange takes place between the two parallel sets of tubes; the system is therefore known as a countercurrent heat exchanger.

In principle, the heat exchange in the nasal passage is similar. In the limb the blood flows in opposite directions in two separate tubes, in the nose the flows in and out take place in the same tube. They are separated in time, however, and we can therefore speak about a *temporal separation* in the countercurrent heat exchanger of the nose, as opposed to the *spatial separation* of the blood vessels in the extremities. This difference is shown diagrammatically in figure 4.

I shall now proceed to discuss the general nature of heat exchange in the respiratory passageways and its importance for the water balance of mammals and birds. The system is not peculiar to desert animals, it is an inevitable consequence of the geometry of the

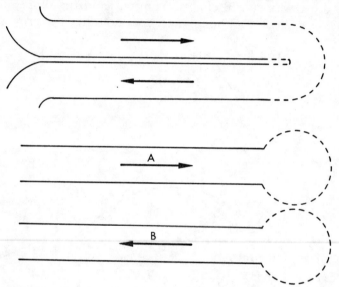

Figure 4. *Top:* Model of countercurrent heat exchanger as in the whale flipper or bird leg. The blood flows in opposite directions in the two tubes, which are separated in space (spatial separation). *Bottom:* Countercurrent heat exchanger as in nasal passageways. On inhalation (A) heat is removed from the wall; on exhalation (B) air flowing in the opposite direction returns heat to the wall. The nasal heat exchanger thus uses one tube in which the two flows are separated in time (temporal separation). (From Schmidt-Nielsen *et al.* 1970*b*.)

passageways, and I shall try to analyze its general features. I shall also show how even cold-blooded animals have something to be gained from cooling of the exhaled air, and finally, I shall show how the type of heat exchanger functioning in the nose of a kangaroo rat can be used to design an engine which appears to be a perfectly feasible *perpetuum mobile*.

Measurements on birds

The observation that air is exhaled, not at body temperature, but at some lower temperature close to the ambient air, can be made on birds as well. We made such measurements on cactus wrens, which are about the same size as kangaroo rats (35 grams), over a range of ambient air temperatures of 12 to 30 °C (see figure 5).

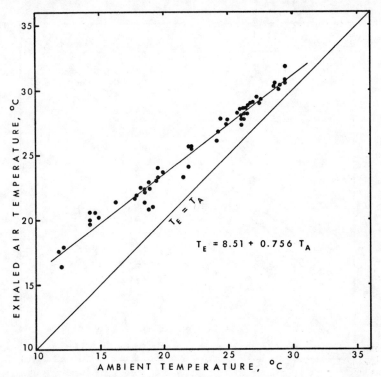

Figure 5. Temperature of the exhaled air of the cactus wren (35 g) at ambient air temperatures between 12 and 30 °C. Body temperature remained constant at 41 °C. (From Schmidt-Nielsen *et al.* 1970*b*.)

The exhaled air was always below body temperature, which was 41 °C, and the extent of cooling ranged from 8 to as much as 23 °C below body temperature (Schmidt-Nielsen, Hainsworth and Murrish, 1970*b*). In other words, over this entire range of ambient temperatures the exhaled air was within a few degrees (1 to 7 °C) of the inhaled air. The cactus wren, like the kangaroo rat, is a desert animal, and cooling of the exhaled air reduces evaporation from the respiratory tract, an obvious advantage in their water balance.

If we compare desert birds with non-desert birds, we might expect to discover some difference between the two groups which may give an advantage in water conservation to the desert species. The results for seven species of birds, which are summarized in figure 6, give no indication that the desert birds are different from

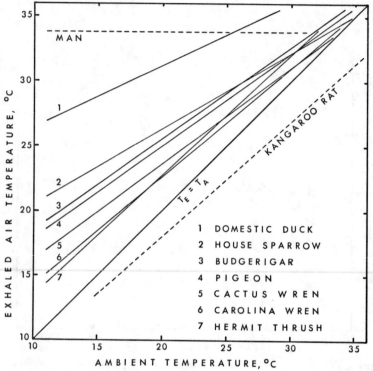

Figure 6. Temperature of exhaled air in seven species of birds. Regression lines obtained from data similar to those in figure 5 and calculated by least squares method. The dashed line for kangaroo rat shows that cooling of the exhaled air is more effective in this animal than in birds. The line labelled *Man* is for one individual; other persons would be represented by somewhat different lines, usually with a positive slope. (From Schmidt-Nielsen *et al.* 1970*b*.)

non-desert birds. In fact, the cactus wren had slightly higher exhaled air temperatures than its close relative, the Carolina wren. Similarly the Australian budgerigar (parakeet), a typical inhabitant of arid lands, was practically indistinguishable from the ordinary house sparrow.

One species of bird distinguished itself by being outside the general range, the domestic duck had considerably higher exhaled air temperatures than the other birds. This is undoubtedly related to the fact that the nasal passages are much wider, and the conditions for heat exchange are therefore far less favorable. The exhaled air temperatures of man would be substantially higher again (although his body temperature is some 3 or 4 degrees below that of the birds), usually in the range of 34 to 35 °C. The reason is the same; the wide passageways are not favorable to heat exchange between air and the surfaces over which it flows.

The kangaroo rat, as we saw before, cools the exhaled air to temperatures below those of the ambient air, and its regression line is therefore located well below the group of birds of similar size. This difference between the birds and the kangaroo rat is easy to understand in terms of the structure of the nasal passageways and the conditions for heat exchange between the respiratory air and the nasal walls. Cross-sections of the nose of a kangaroo rat (figure 7) show a very narrow passage with a large wall surface;

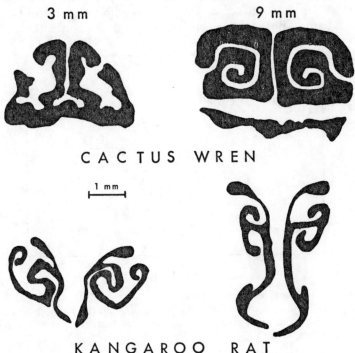

Figure 7. Cross-sections of the nasal passageways of cactus wren and kangaroo rat. The passages are wider and the wall area smaller in the bird than in a mammal of the same body size (about 35 g). In both animals the profiles were obtained at a depth of 3 mm and 9 mm, respectively, from the external openings. (From Schmidt-Nielsen *et al.* 1970*b*.)

this facilitates to the utmost exchange of heat between the air and the nasal tissues as it flows closely over the surface. In the birds, where the passageway is wider and shorter, the surface area is smaller, and conditions for heat exchange are less favorable. In the duck the passageways are larger again, and in man the heat exchange between air and nasal walls is so incomplete that under most circumstances it is of only minor importance. Thus, the statement, that air is exhaled saturated at body temperature or very nearly so, is quite true for man and is in accord with actual observations. The mistake has been to generalize by assuming that this would hold true for all other warm-blooded animals.

Effects on water balance

If we wish to examine the importance of cooling of the exhaled air for the water loss from an animal, it is helpful first to recall that the water vapor content of saturated air is highly dependent on the temperature. The curve shown in figure 8 illustrates the rapid

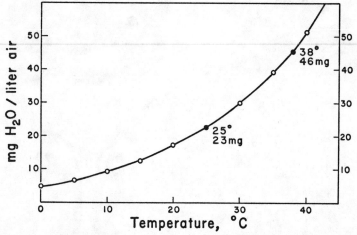

Figure 8. Water vapor content in air at saturation (100% relative humidity). The slope of the curve becomes increasingly steep around the body temperature of birds and mammals, thus greatly increasing the amount of water required for saturation of lung air.

increase in the amount of water vapor that can be held in air at higher temperatures. As a rule of thumb, I have found it convenient to remember that saturated air at body temperature contains exactly twice as much water vapor (46 mg l⁻¹) as saturated air at room temperature (23 mg l⁻¹). Cooling of the exhaled air from body temperature to room temperature therefore reduces by one-half the amount of water vapor that the air can carry away, indeed a substantial water saving when we remember that evaporation is responsible for the major water loss from the animals we have under consideration.

The effect of cooling the exhaled air on the water balance of the kangaroo rat is shown in figure 9, and this graph can be used

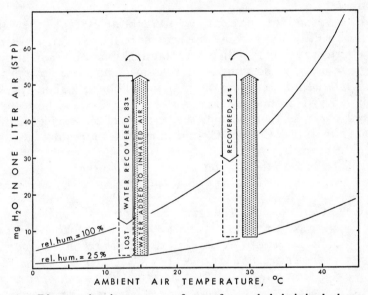

Figure 9. Diagram showing recovery of water from exhaled air in the kangaroo rat. When air is inhaled at 15 °C and 25% relative humidity, the amount of water needed to bring this air to saturation at body temperature (38 °C) is indicated by the shaded bar placed at 15 °C on the abscissa. Under these atmospheric conditions the kangaroo rat exhales air at 13 °C, and the amount of recondensed water is represented by the unshaded column. The bars to the right in the diagram give evaporation and recovery at ambient air of 30 °C, 25% relative humidity. (From Schmidt-Nielsen *et al.* 1970*b*.)

as an illustration of the effect on small rodents in general. These calculations are for two arbitrarily selected air temperatures, 15 and 30 °C, both at the equally arbitrarily chosen 25% relative humidity. In this diagram the shaded bars indicate the amount of water vapor that must be added to one liter of outside air as it is inhaled and brought to saturation at body temperature (38 °C).

The unshaded bar which is placed next to each shaded bar is located at the corresponding temperature of exhaled air under the selected conditions. If air is inhaled at 15 °C, for example, exhaled air will be at 13 °C. In this case the cooling of the exhaled air causes the recondensation of 83% of the water that was added on inhalation, i.e. only about one-sixth of the water needed to saturate the respiratory air is lost by the animal. At 30 °C the savings are smaller, but even so, 54% of the evaporated water is saved as compared to what would be lost had the air been exhaled at body temperature.

In a typical bird, the cactus wren can be an example, the water savings will be less because the cooling of the exhaled air is not as effective, yet they will be substantial. For conditions similar to those used for the kangaroo rat, an air temperature of 15° and 25% relative humidity, the water saving will be 74%, and at 30° it will be 49% (Schmidt-Nielsen *et al.* 1970*b*). It should be remembered, of course, that the body temperature of birds is higher than in mammals (41 °C *v.* 38 °C) and more water is required to saturate the lung air.

What effect do these water savings have on the heat budget of these small animals? Table 3 shows an account of the heat that a

Table 3. *Comparison, for the kangaroo rat, of the amount of heat required for warming and humidification of one liter of inhaled air with that recovered by cooling of the same air on exhalation. Under the chosen atmospheric conditions (15 °C, 25% relative humidity, as in figure 9, left), 88% of the added heat is recovered. The recovered heat is 16.1% of the metabolic heat produced concurrently with the breathing of 1 liter of air (192 cal.).*

	Added upon inhalation	Recovered on exhalation	% Savings on exhalation
Heat of vaporization	28.5 cal	23.5 cal	83%
Heating of air	6.8 cal	7.4 cal	107%
Total heat	35.1 cal	30.9 cal	88%

Savings in % of metabolic heat production = 16.1%

kangaroo rat adds to the air during inhalation, and how much is recovered again on exhalation. It turns out that, at 15 °C, 35 calories are used to heat and humidify one liter of respired air, but

of this heat 31 calories are recovered on exhalation. Thus, some 88%
of the heat used to warm and humidify exhaled air is recovered
again, and therefore does not constitute a drain on the body heat.
It is evident that this is quite significant, for it is 16% of the simultaneous metabolic heat production. At higher ambient temperatures the relative importance of heat savings will be less, but at
lower temperatures it will be greater.

In man the large dimension of the nasal passage and a laminar
air stream cause a far less complete heat exchange between the
stream of respired air and the nasal walls, the distance from the
center of the stream to the wall is too great. Man exhales air at
close to body temperature, and the recovery of heat and water
vapor is usually insignificant. At very low air temperatures, however, the amount of heat that is lost due to heating and humidification of inhaled air becomes a substantial item in the heat budget
even in man. For example, for a man breathing dry air at 0 °C,
nearly one-fifth of his metabolic heat is used to heat and humidify
the respiratory air. At even lower air temperatures the passageways begin to cool off substantially, for example at −30 °C,
recovery of heat and water from the exhaled air becomes quite
important in reducing the drain on the heat resources of a man
(Walker, Wells and Merrill, 1961; Webb, 1951, 1955).

Mathematical model of nasal heat exchange

If we wish to examine the nasal passageway as a heat exchanger,
we find that there is a formidable number of variables to be considered. As heat diffuses between air and wall, water is either
evaporated or condensed, and in the process binding or releasing
substantial amounts of latent heat. Of obvious importance for heat
and water flux are such geometrical variables as the total surface
available for heat exchange, and the distance from the core of the
air stream to the wall. External physical conditions, air temperature and humidity, must be important, for both influence heat
transfer and evaporation. We also have a number of physiological
variables to consider; the rate of respiration and the volume of
respired air can vary, and both influence the linear velocity of the
air stream flowing over the nasal surfaces, a velocity which is
continuously changing during each respiratory cycle. Then there
is the blood stream that not only supplies the water that is evaporated, but also carries heat from the body core to the nasal region.

During inhalation heat flows transversely from the wall to the air stream, and water vapor flows into the air along the transversal temperature gradient thus established. On exhalation this process is reversed, heat flows from the air to the nasal wall, and, as the temperature decreases, water vapor diffuses transversally and recondenses on the surface. To describe these processes adequately during the constantly changing linear flow rates during a single breath has proved an unmanageable problem.

One of my former students, Dr J. Collins, has shown that this intractable number of variables can be disregarded and the nasal passage can be analyzed as a relatively simple steady-state model. Figure 10 shows in diagrammatic form a model of the nasal passage,

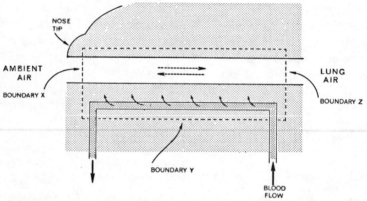

Figure 10. Heat and water exchange in the nasal passageway can be analyzed as a steady-state model by examining the energy transfer across an imaginary boundary indicated by the dashed line in this diagram. A mathematical model, as described in the text, can be used to predict nose temperatures and efficiency of water recovery as given in tables 4 and 5.

a liquid flow stream (blood), and an imaginary boundary indicated by a dotted line. This imaginary boundary encloses the nasal passage and its moist surface, beginning at a given distance from the the nose tip at Boundary X, and extending along Boundary Y to Boundary Z where the temperature equals body-core temperature. Energy transported across this boundary can be separated conceptually into several components. First of all, the air crossing Boundary Z undergoes no change in energy content, for by definition this boundary is where the air is at body temperature and saturated. The energy crossing Boundaries X and Y can be separated into four components:

(a) change in air temperature,
(b) evaporation of water,
(c) thermal conduction to outside,
(d) change in blood temperature.

This permits the development of a mathematical model which describes the steady-state energy balance of the system, in which the rate of energy transfer from the liquid stream (designated (d) above, representing blood flow and heat diffusion from the body core to the nasal region) is equated with the sum of energy transfer (a) by a change in the temperature of respiratory air, (b) by evaporation, and (c) by direct conduction from the nose tip itself to the environment, or $(a) + (b) + (c) = (d)$. All the variables which enter the equation, except the flow rate of the liquid stream (d), can be measured or estimated from physiological measurements. This permits the solution of the equation for the liquid stream flow rate. After having obtained an empirical solution for the liquid stream flow rate, this variable can be eliminated and a general energy balance equation can be completed, which in turn can be used to compute values for expired air temperatures, rates of water loss, and efficiency of vapor recovery, for a variety of ambient air conditions.

I shall omit a discussion of how the mathematical model was justified and tested against physiological measurements, for the lengthy details have been given elsewhere (Collins, Pilkington and Schmidt-Nielsen, 1971). Instead I give two tables which show the computer print-out for predicted nose temperatures of kangaroo rats under a variety of atmospheric conditions, and the efficiency of predicted water recovery in the nose. Table 4 shows model-predicted nasal temperatures. It contains no great surprises, the exhaled air temperature depends very much on ambient conditions, and especially the relative humidity of the inhaled air is important. At high humidities, cooling of the exhaled air, as expected, is uniformly not as great as in dry air.

More surprising is the model-predicted efficiency of water recovery in the nasal passages (table 5). Under the wide variety of ambient air temperature from 10 to 36 °C, and relative humidities from 0 to 100%, the predicted efficiency of water recovery varied only from 66 to 77%. In other words, in spite of changing ambient conditions the model predicts that the efficiency of water recovery should remain fairly close to three-quarters, a finding that without the aid of a computer would require an inordinate amount of

Table 4. *Computer-predicted nose tip temperature of the kangaroo rat as a function of ambient conditions.* (From Collins *et al.* 1971)

Amb. temp. (°C)	Ambient relative humidity (%)					
	0	20	40	60	80	100
10	9.8	11.8	13.7	15.3	16.8	18.1
12	10.7	12.9	14.8	16.5	18.1	19.5
14	11.7	14.0	16.0	17.8	19.4	20.9
16	12.6	15.1	17.2	19.1	20.8	22.4
18	13.6	16.2	18.4	20.4	22.2	23.9
20	14.6	17.3	19.7	21.7	23.6	25.3
22	15.6	18.5	21.0	23.1	25.1	26.9
24	16.6	19.6	22.2	24.5	26.5	28.4
26	17.7	20.8	23.5	25.9	28.0	29.9
28	18.8	22.0	24.9	27.3	29.5	31.4
30	19.7	23.2	26.1	28.6	30.9	33.0
32	20.6	24.3	27.4	30.0	32.4	34.5
34	21.6	25.4	28.6	31.4	33.8	36.0
36	22.6	26.6	29.9	32.8	35.3	37.6

Table 5. *Computer-predicted efficiency of nasal vapor recovery in the kangaroo rat as a function of ambient conditions.* (From Collins *et al.* 1971)

Amb. temp. (°C)	Ambient relative humidity (%)					
	0	20	40	60	80	100
10	0.755	0.756	0.758	0.760	0.763	0.766
12	0.749	0.750	0.753	0.755	0.758	0.762
14	0.743	0.745	0.747	0.750	0.754	0.757
16	0.736	0.739	0.741	0.745	0.749	0.753
18	0.730	0.732	0.735	0.739	0.743	0.748
20	0.723	0.725	0.729	0.733	0.737	0.743
22	0.715	0.718	0.722	0.726	0.731	0.737
24	0.707	0.710	0.714	0.719	0.725	0.731
26	0.698	0.702	0.706	0.711	0.717	0.724
28	0.688	0.692	0.697	0.703	0.709	0.715
30	0.682	0.686	0.692	0.697	0.703	0.710
32	0.675	0.680	0.685	0.691	0.697	0.704
34	0.667	0.672	0.678	0.684	0.690	0.696
36	0.659	0.665	0.671	0.677	0.682	0.686

repetitive and rather dull laboratory observations. Although the model has a margin of uncertainty in its predictions, we feel that there is some merit in asking a computer to do this work and instead turn our attention to other problems.

'Cold-blooded' animals

If we measure the exhaled air temperature in a lizard, it will be very nearly the same as the body temperature (figure 11). This

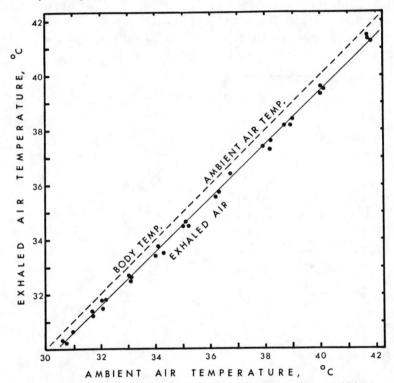

Figure 11. Temperature of the exhaled air of the desert iguana (*Dipsosaurus dorsalis*). Body temperature is nearly equal to ambient air temperature; the exhaled air is slightly cooler than the body due to evaporation in the respiratory passages. (From Murrish and Schmidt-Nielsen, 1970.)

is no surprise, for body and ambient air temperatures are near-equal, and the slight cooling of the exhaled air which is observed is due to the fact that the surfaces of the respiratory system are moist and somewhat cooled by evaporation. So far, everything is as expected, and the evaporation from the respiratory tract will, besides respiratory volume, depend only on ambient temperature and humidity.

A complication arises, however, from the well-known fact that the body temperature of lizards by no means always equals that of the surrounding air; many lizards regulate their body temperature at a level which may be far above that of the surroundings, mainly by basking in the sun. Since the respiratory air is saturated with water vapor in the lungs, the high body temperature means that significantly larger amounts of water are now used to saturate the respiratory air. The lung air is at body-core temperature also when body temperature and surroundings differ, but the air is not necessarily exhaled at this temperature.

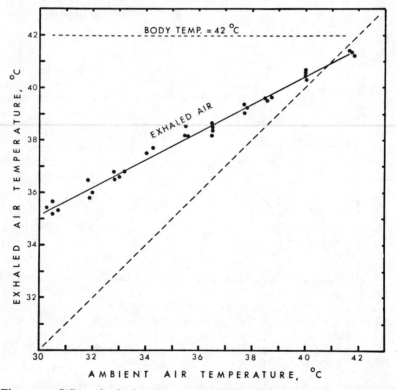

Figure 12. When the body temperature of the desert iguana is maintained constant at 42 °C, the exhaled air will be lower than the body temperature, although the air in the lung is at 42 °C. The extent of cooling of the air varies with the ambient air temperature. (From Murrish and Schmidt-Nielsen, 1970.)

The desert iguana (*Dipsosaurus dorsalis*) in nature is frequently found to have a body temperature of about 42 °C, even when the air temperature is as low as 30 °C (Norris, 1953). This it achieves by exposing itself to the sun, even to the extent of selecting the most favorable posture for intercepting the maximum amount of solar radiation. In the laboratory we maintained the body temperature of these lizards at a constant 42 °C by means of simulated solar radiation while measuring the temperature of the exhaled air at a variety of ambient air temperatures. Cooling of the exhaled air was found to be considerable (figure 12). For example, the exhaled air was 7 °C below body temperature when the latter was 42 °C and the room air was 30 °C (Murrish and Schmidt-Nielsen, 1970). These particular conditions could be a normal situation for a desert iguana while basking in the desert sun in the early morning, and the observed cooling of the exhaled air would reduce the respiratory water loss by 31% relative to what the loss would have been had the air been exhaled at lung temperature.

The principle which underlies the cooling of the exhaled air is the same as that in mammals and birds, but the efficiency of exchange is much lower, and hence the water savings as well. This is due primarily to the different shape of the passageways; they are wider and shorter and do not have the large total surface that we found in the kangaroo rat.

The geometry of the nasal passageway of the desert iguana has one characteristic, however, that leads to some very interesting consequences. The most distal part of the passage, just inside the external nares, forms a slight depression (figure 13A). Fluid secreted from the nasal salt gland (located at B and C) accumulates in this depression and contributes to the humidification of inhaled air. As water is evaporated, this fluid gradually becomes more concentrated, and salts crystallize and form encrustations in or around the openings of the nares. This means that water contained in the secretion from the nasal gland does not constitute a net loss to the body. Since water must in any event be used to humidify the respiratory air, evaporation from the nasal fluid serves to utilize water that has already been 'eliminated' as far as body water is concerned. Thus the utilization of this water for humidification of respiratory air has the same net effect on the water balance of the body as if the nasal gland were able to secrete the salts in crystalline state, without the use of any water.

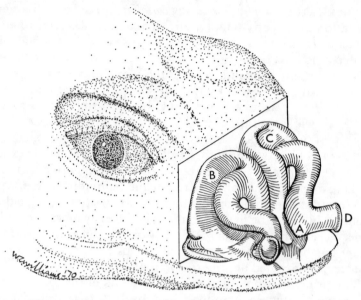

Figure 13. The nasal passage of the desert iguana forms a slight depression (A), just inside the external nares (D). Fluid from the nasal salt-excreting glands (located at B and C) accumulates in this depression and contributes moisture to the humidification of the respiratory air. (From Murrish and Schmidt-Nielsen, 1970.)

A *perpetuum mobile?* or
A little-known heat engine

The diagram in figure 14 shows an engine which utilizes heat exchange in a way similar to that in the nose of the kangaroo rat. The engine consists of two interconnected air-tight cylinders. The smaller cylinder contains an ordinary air-tight piston, but the piston in the larger cylinder is of a special nature. To reduce friction it is not quite in contact with the cylinder walls, and the piston itself permits passage of air through its body between closely packed vertical plates of a suitable material. One end of the large cylinder is kept hot, the other is cool. If, as indicated in the diagram, the large plunger is moved downwards, which can be done with little expenditure of energy because of the loose fit and easy passage of air through the piston itself, the heated air in the lower part of the cylinder will give off heat to the plates in the piston. If the heat exchange is effective, air will exit at the upper surface of the piston cooled to piston temperature. When the large piston is in the bottom position, the entire system will be filled with cold air, and the reduced pressure forces the piston in the small cylinder upwards.

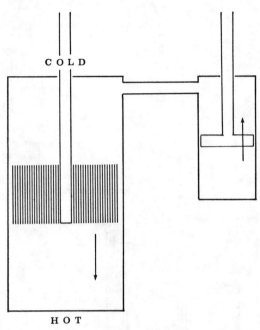

Figure 14. This diagram shows a heat engine which at first glance seems to work as a *perpetuum mobile*. It is based on heat exchange in the large piston, similar to the heat exchange in the kangaroo rat nose. Details of its function are given in the text.

The next stage of the cycle consists of the reversal of the direction of the large piston, cold air flowing over the heated plates of the piston picks up their heat, and in the uppermost position the entire large cylinder is filled with hot air. The increased pressure therefore forces the small piston down, thus completing the cycle. Since the large piston moves with little friction and air flows freely through its body, only a small amount of power is necessary to move it up and down. This power requirement is taken off as a fraction of the power available from the small piston, the remainder being available for external work.

The functioning of this engine depends on the heat exchange in the large piston. Heat is added to air, and again removed from it as the direction of flow is reversed in the exchanger. At first glance, it seems that the engine should be a *perpetuum mobile*. Theoretically, the frictional losses in the large piston could approach zero, and heat exchange could theoretically approach 100%. On second thought, it may be that a *perpetuum mobile* still remains impossible, and that the whole scheme is based on some incorrect assumption.

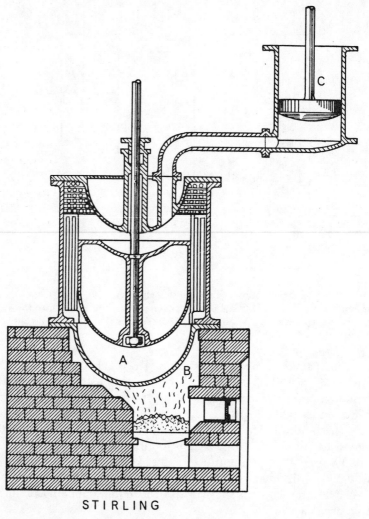

STIRLING

Figure 15. The Stirling heat engine, patented in 1827 by the Reverend Robert Stirling, works on the principle described in figure 14. (From Jenkin, 1885.)

I can assure you, however, that this is not the case. This heat engine was invented by a Scottish minister, the Reverend Robert Stirling, and was patented in 1827, and again in a revised form in 1840. It was not only patented, but was built and used as well (figure 15). Engineering literature from the last century reports on a 50 horsepower Stirling heat engine which was used for many years in a Dundee foundry. The thermodynamic efficiency of the Stirling engine has been examined and is limited by the same considerations as the Carnot heat engine; the cycles are in fact identical, except that the Carnot engine operates with adiabatic conditions as opposed to the constant-volume operation of the Stirling engine.

I should like to conclude this discussion by noting that the use of heat exchangers or heat regenerators of the Stirling type are being increasingly used in a variety of applications. One such use is for attaining very low temperatures, below those of liquid helium (Gifford and McMahon, 1960). Temperatures approaching absolute zero are easily reached with heat exchangers which utilize the same principle as that which is the single most important factor in the water balance of small rodents in some of the hottest deserts of the world.

2. Panting and heat loss

In the preceding chapter my main emphasis was on how difficult it is to lose heat and water from the respiratory tract, especially for small animals. Nevertheless, many animals, both mammals and birds, resort to evaporation from the respiratory tract to dissipate excess body heat. This is the phenomenon we know so well as panting in dogs, and for that matter in other domestic animals as well, such as sheep and cattle. Birds also respond to a heat load with a greatly increased respiratory activity, although there are differences from the mammalian type of panting to which I shall return later.

On cooling hot dogs

The panting of dogs is often described as a rapid in-and-out breathing through the open mouth with evaporation taking place from the moist oral surfaces and the large hanging tongue.* Air flowing close to the moist surface will rapidly become saturated with water vapor, but much of the air will not be in the immediate vicinity of the moist surface and it is easy to imagine that the entire volume of air does not readily become fully saturated. In that event correspondingly larger volumes of air would have to be moved, thus increasing the muscular effort of breathing, and the heat load due to the increased energy expenditure.

A priori it seems that the moist surfaces of the nose are better suited for evaporation, for here the air flows closely over the large moist surfaces of the nasal turbinates and rapidly attains near-complete equilibrium with respect to both temperature and water vapor (Verzár, Keith and Parchet, 1953). The difficulty in using the nose for heat dissipation is that, on exhalation, much of the

* Many mammals pant through the nose with closed mouth, and only during the most severe heat loads when they are visibly distressed do they open the mouth and change to a slower, deep and labored respiration.

heat and water vapor that was added to the inhaled air is again recovered through countercurrent exchange on exhalation. One way to circumvent this recovery of heat is to arrange for a one-way flow of air in the nose and use the mouth for exhalation.

During shallow thermal panting in dogs we have found that the typical air flow is indeed unidirectional, with inhalation predominantly through the nose and exhalation through the mouth (figure 16). In the dogs we tested, on the average about a quarter of the air inhaled through the nose was exhaled again through the nose, the remaining three-quarters being exhaled through the mouth. The amounts could vary a great deal, however, and at any given moment from zero to 100% of the inhaled air volume could be exhaled through the nose (Schmidt-Nielsen, Bretz and Taylor, 1970a).

The described pattern of air flow in the panting dog means that the nasal mucosa, rather than the mouth and the tongue, is the primary site of evaporation. Consequently, the nose must be supplied with sufficient quantities of moisture, and we assume that the large nasal gland which is found in the dog is the major source of the necessary water. The nasal gland of the dog was described by the Danish anatomist Nicolaus Steno in 1664, but an analogous gland is not found in man. For this reason its existence is not as widely known as otherwise it would be. No specific function, except to 'keep the nose moist', has been ascribed to this gland, but if indeed it supplies the water for heat dissipation during panting, its function is in a sense analogous to that of the sweat glands in man. It would therefore be of interest to examine whether the secretory activity of this gland is under control of the thermoregulatory system in the same way as sweat glands, and I would encourage such studies.

The importance of a unidirectional flow of air in the nose can be estimated from the temperature of the exhaled air. When the heat load is moderate, a dog may pant with closed mouth, both inhalation and exhalation taking place through the nose. Under these circumstances we found the temperature of exhaled air to be 29 °C. If the same dog, while still panting at the same frequency, changed to exhalation through the mouth, the temperature of the exhaled air was nearly identical to body temperature, 38 °C. If air is exhaled at a higher temperature, it carries with it more heat and more water, thus increasing heat dissipation. If one liter of air is exhaled saturated at 38 °C instead of 29 °C, it carries away 27.7 cal instead of 14.9 cal. Exhalation through the mouth thus nearly doubles the heat loss. By changing the relative amounts of air exhaled via the nose and mouth, the dog can therefore modulate the amount of heat dissipated, without changing the frequency of respiration or the tidal volume.

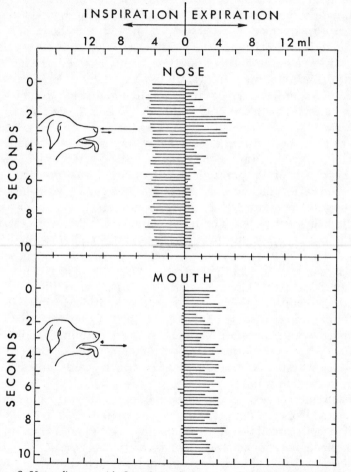

Figure 16. *Upper diagram:* Air flow through the nose of a panting dog. Horizontal lines extending to the left of the vertical mid-line indicate inhalation, to the right, exhalation. Mean inhaled and exhaled volumes are indicated by vectors placed adjacent to the dog nose. *Lower diagram:* Air flow through the mouth of a panting dog. Symbols as above. Inhalation through the mouth is virtually zero, exhalation through the mouth carries most of the air taken in through the nose. (From Schmidt-Nielsen *et al.* 1970a.)

This brings us to the interesting fact that dogs, and many other animals, seem to pant at a relatively constant frequency. When panting begins, respiration suddenly changes from, say, 30 or 40 respirations per minute to a relatively constant high level between about 300 and 400. A dog subjected to a very light heat load will not pant at intermediate frequencies; instead, brief periods of panting at the high frequency will alternate with periods of normal slow respiration.

About ten years ago it was suggested by Eugene Crawford, then a student at Duke University, that dogs seem to pant at a resonant frequency (Crawford, 1962). Due to its elastic properties, the entire respiratory system has a natural frequency of oscillation, and to keep the system oscillating at this frequency requires the expenditure of only minimal muscular effort. As a consequence, the heat production by the respiratory muscles is at a minimum, thus adding only little to the heat load. Crawford estimated that if panting were to take place without the benefit of a resonant system, the increased muscular effort would generate more heat than the total amount the panting process can dissipate.

If panting takes place at a constant resonant frequency, a modulation of the amount of heat dissipated must be achieved without changing this frequency,* and we now see the value of modulating heat loss by merely changing the flow of air through the nose and the mouth.

A well-regulated modulation of flow through the nose permits precise modulation of heat loss: (1) without changing frequency, which energetically would be expensive and produce additional heat in the muscles, and (2) without changing depth of breathing, which would have profound consequences for gas exchange in the lungs.

Panting at moderate heat loads does not cause overventilation of the lung (Albers, 1961), but severe heat load causes the respiratory minute volume to be increased, with overventilation gradually causing excessive loss of carbon dioxide from the blood and leading to alkalosis. Hales and Bligh (1969) observed that the respiratory frequency of dogs immediately upon heat exposure increased 18-fold, from 18 to 340 breaths/min. With increasing heat loads respiratory frequency rose to a peak rate of 410, but ultimately declined again to 260. During this second phase the

* Some change in panting frequency can and does occur; changes in muscle tonus will change the natural frequency of the visco-elastic oscillating system.

respiratory minute volume and alveolar ventilation were greatly increased and respiratory alkalosis developed.

Separation of panting into two phases has also been observed in sheep and cattle (Hales and Findlay, 1968). During the first phase the dead-space ventilation in the ox can be increased many-fold without any appreciable change in alveolar ventilation (figure 17), and only during severe heat stress, when breathing becomes labored and open-mouthed and the animal is in apparent distress, does alveolar ventilation begin to increase. In this second stage the total ventilation continues to increase, and although the respiratory frequency at the same time declines, the animal now becomes severely alkalotic.

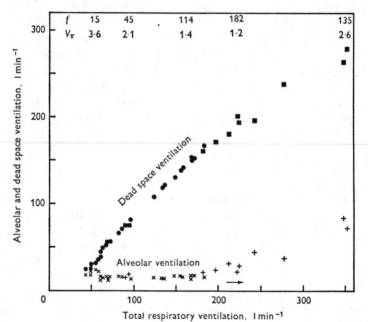

Figure 17. As the total respiratory ventilation (abscissa) increases in the panting ox, the dead space ventilation increases steadily. The alveolar ventilation, however, does not increase until the total ventilation exceeds about 200 liters per minute. In extreme panting the respiratory frequency (f) is decreased as tidal volume (V_T) is increased (figures at top of graph). (From Hales, 1966.)

Panting birds

Most birds will pant if they are subjected to a heat load. Such panting is an easily observable increase in respiration, and some birds supplement this with a rapid gular fluttering. In discussions of bird respiration and cooling at high air temperatures, the distinction between panting and gular flutter has not always been clear. Panting, as I will use it here, refers to breathing movements of the thorax and abdomen, while gular flutter is the rapid movement of the thin floor of the buccal cavity and the upper region of the throat. I shall give little attention to gular flutter, and instead concentrate on panting. The phenomenon of panting is particularly interesting in birds, for the avian respiratory system in some respects differs radically from that of mammals.

Briefly, the respiratory system of birds consists of a pair of lungs where gas exchange with the blood occurs, and of several large, thin-walled air-sacs that act as bellows in the movement of gases over the exchange surfaces of the lungs. The air-sacs are poorly vascularized and do not participate directly in gas exchange between air and blood. This has been shown by leading carbon monoxide into an air-sac whose connections to the other parts of the respiratory system had been plugged; the carbon monoxide is not taken up by the blood in sufficient quantity to poison the bird.

While the mammalian lung is ventilated by an in-and-out flow of air, the bird lung will permit a through-flow of air. The finest branches of the bronchi do not end in sacs like the mammalian alveoli, they are tubes which are open at both ends, and gas exchange takes place in air capillaries which surround these bronchi, known as parabronchi or tertiary bronchi (figure 18).

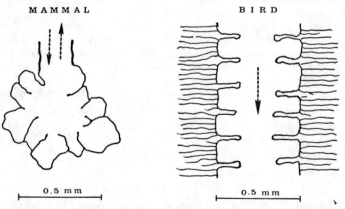

MAMMAL   BIRD

0.5 mm   0.5 mm

Figure 18. The smallest units of the lung of mammals end in blind sacs, alveoli. In birds the finest branches of the bronchi (the parabronchi) are tubes which are open at both ends and permit through-flow of air.

There is a large body of anatomical information on the avian respiratory system, but the question of how air flows in the various passageways of the bird lung has remained uncertain, although the subject of numerous hypotheses. Many excellent reviews have been published; one of the clearest and most comprehensive being a recent contribution by A. S. King (1966).

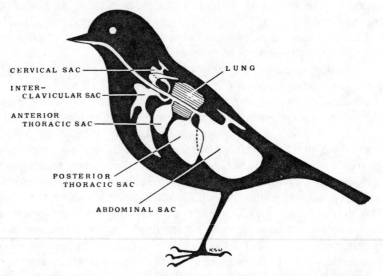

Figure 19. The body of a bird is filled with several large, thin-walled air-sacs which extend in between other organs, ramify, and connect with air spaces in the hollow bones. The paired lungs are relatively small, and each is supplied by a main bronchus (mesobronchus) which continues through the lung to the posterior air-sac. In the anterior part of the lung the main bronchus connects with the three sets of anterior air-sacs. The lung itself has connections not only with the main bronchus, but also direct connections (called recurrent bronchi) with some of the air-sacs. (Modified after Salt, 1964.)

Many of my readers may be quite familiar with this subject but I still wish to introduce some points that are particularly pertinent. The complex air system fills some 15% of the body volume of birds, the lungs are fairly small and rigid, and most of the volume is in the air-sacs that also connect with pneumatized spaces in the bones and elsewhere (figure 19).

The most peculiar aspect of the anatomy is the complexity of air passages which connect the various parts. A large main bronchus (called the mesobronchus) runs directly through the lung, sends off branches to the anterior air-sacs, connects to a system of fine branches in the lung parenchyma, and continues directly to the posterior air-sacs. The air-sacs also connect to the lung itself through passages known as recurrent bronchi, thus further increasing the multitude of channels and directions in which air could flow.

In diagram form these main connections can be represented as shown in figure 20. In this diagram I have combined all the air-sacs

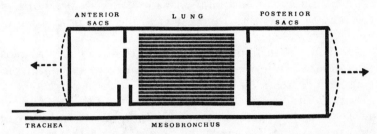

Figure 20. In diagram form the complex respiratory system of birds can be reduced to a small number of functional units: the anterior air-sacs combined in one space, the posterior in another; the lung itself built so that it permits through-flow of air and with connections both to the mesobronchus and to the air-sacs. On inhalation, the air-sacs increase in volume, as indicated by dotted lines and arrows.

that are located in the anterior part of the bird into one space, and all those located posteriorly into another, the lung being located in between. The connections from the air-sacs directly to the lung are also shown. The most significant aspect of having a direct connection between the large mesobronchus and the air-sacs is that this seems to permit an increased ventilation of the air-sacs alone without involving the lung itself. In other words, there is a bypass or shunt which during panting might permit an increased flow of air, yet without passing an increased volume of air through the lung itself, a most desirable arrangement that could serve to prevent alkalosis.

One observation that is important to our further discussion is that many birds seem to pant at a rather constant frequency. It has been suggested that this is due to the existence of a resonant frequency, similar to that which Crawford demonstrated for dogs. For example, when a pigeon begins to pant, the respiratory frequency changes from 30 cycles/min to about 600 cycles/min.* Recently, Crawford and Kampe (1971) have demonstrated that the resonant

* During work on heat-stressed pigeons in my laboratory Dr Lutz has found more variable panting frequencies than Crawford reports. At the present time we are not aware of the reasons for these differences.

frequency of the respiratory system of pigeons is indeed practically identical to the normal panting rates (resonant frequency, 564 cycles/min; panting frequency, 612 cycles/min). The work done to overcome the forces opposing movement of the respiratory system are therefore at a minimum at this frequency. This is an important finding, for it shows that the panting bird can provide sufficient ventilation to augment evaporation with a small expenditure of energy (and thus a minimal increment in the heat production of the respiratory muscles).

It is necessary to mention, however, that this by no means holds true for all birds. In cormorants, for example, there is a continuous and uniform increase in respiratory rates with increasing body temperature, although the rate of gular flutter in these birds always takes place at a high and constant frequency, highly suggestive of an oscillating system (Lasiewski and Snyder, 1969). In cormorants and pelicans the breathing and flutter rates are quite independent; in the pigeon, however, panting and flutter are completely synchronized (Calder and Schmidt-Nielsen, 1966). This also holds true for many other birds: goose, duck, chicken, gull, roadrunner, and others. It is worth noting that some birds pant without showing any signs of gular flutter; this seems to hold for vultures and all passerine birds.

Modulation of heat dissipation

In a bird where panting frequency is determined by the resonant characteristics of the respiratory system, frequency remains constant and other means must be found to achieve a modulation in evaporation and heat dissipation. The simplest, at moderate heat loads, is to intersperse brief periods of panting with periods of normal respiration, just as dogs do. Many birds do indeed behave this way; at moderate heat loads panting takes place intermittently. At more severe heat loads the birds pant continuously, and further modulation must be achieved by increasing the total flow of air (the respiratory minute volume). In this situation we would expect the anatomical shunt to be effective in bypassing the lung, thus eliminating the severe alkalosis that is characteristic of heavily panting mammals.

To our surprise we found that pigeons when they pant vigorously develop severe alkalosis. This indicates that shunting, if it exists at all, is insufficient to prevent overventilation of the lung (Calder and Schmidt-Nielsen, 1966). Further study of several other bird species showed much the same pattern; panting is associated with a decrease in the CO_2 pressure in the blood and an alkaline shift of the blood pH to a degree that would be lethal to man (figure 21).

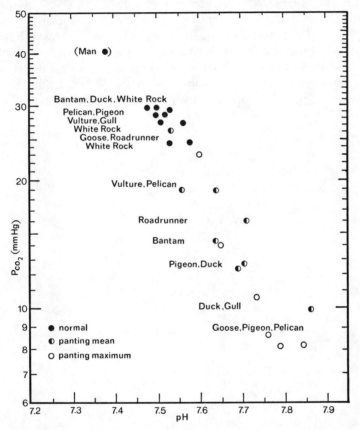

Figure 21. The CO_2 tension in arterial blood of birds at rest (filled circles) is about 30 mm Hg or slightly less. This is substantially lower than in mammals, which have about 40 mm CO_2. Birds that pant (open, divided circles) have lower CO_2 tension, and those that pant maximally (empty circles) may have CO_2 tensions as low as 8 mm Hg and a blood pH of 7.9. All points represent mean values for each species. (From Calder and Schmidt-Nielsen, 1968.)

It would be a mistake, however, to conclude that all birds become severely alkalotic during panting. The ostrich, which is native to hot, dry areas of Africa and the Near East, is highly tolerant to heat loads and has, in our experiments, withstood an air temperature of 56 °C for as long as 5 hours while panting constantly at a frequency of 50 cycles/min. In a panting ostrich the respiratory volume may be increased as much as 20-fold compared to the resting bird. The gas composition in the air-sacs now becomes more similar to outside air with a CO_2 concentration of less than 1 %. This shows that the sacs are heavily ventilated, and yet no alkalosis develops (Schmidt-Nielsen *et al.* 1969). In the ostrich, then, it seems that the anatomical shunt is fully operational, and why this differs in other birds remains a mystery. If it is related to the fact that the ostrich is a non-flying bird we do not understand how, for its air-sac system is highly developed and similar to that of other birds.

3. How birds breathe

How air flows in the bird lung

The finding, much against expectation, that most birds seem not to use the obvious anatomical shunt to bypass the lung when they pant, makes it even more desirable to clarify the patterns of air flow in their complex respiratory system.

The respiratory system of mammals is simple; the sac-like lungs and their subdivisions are ventilated by an in-and-out flow of air with diffusion playing the major role at the terminal units, the alveoli. In the bird lung, however, there is no need for a reversal of the air flow because the finest branches of the bronchi, the parabronchi (tertiary bronchi), are open at both ends and permit through-flow of air. This characteristic, in combination with the multiplicity of tubes, branches and connections to the air-sacs, has led to a great deal of speculation as to how air flows in this complex system. Virtually all imaginable possibilities have been suggested, including some that violate physical laws, observable fact, and physiological evidence. I should therefore like to summarize the rather strong evidence we now have that air flows unidirectionally through the lung both during inhalation and exhalation, and then discuss the consequences that this has for the function of the avian lung as an efficient organ for oxygen uptake. Although bird blood has no greater affinity for oxygen than mammalian blood, the arrangement of the avian lung permits a far better utilization of oxygen than in mammals at low atmospheric pressure, which is of importance to bird flight at high altitude.

It was during the work on the ostrich that the need for further understanding of bird respiration became acute, for our results suggested unidirectional flow in the lung in the cranial direction, during both inhalation and exhalation.

It is interesting that the ostrich, although it is a non-flying bird, has a well-developed air-sac system with the same major structures as in other birds and a total volume which is about 15% of the body volume (Schmidt-Nielsen *et al.* 1969). In order to gain some information about air flow in this system, we made an ostrich inhale a single breath of pure oxygen as a tracer gas, and then

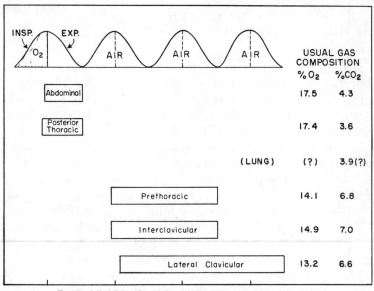

TIME OF ARRIVAL OF INHALED OXYGEN RELATIVE
TO BREATHING CYCLE (TOP OF GRAPH)

Figure 22. When an ostrich inhales a single breath of pure oxygen, the arrival time of the oxygen at various points can be followed with an oxygen electrode. The respiratory cycle is indicated at the top of the graph, the shaded portion of the first cycle denotes one single inhalation of pure oxygen. Respiratory rate about 6 cycles min, i.e. each cycle about 10 seconds. The left end of each horizontal bar indicates the earliest recorded arrival time in that particular sac; the right end of the bar is the latest time at which the oxygen concentration was observed to increase. The oxygen arrives at the posterior air-sacs earlier than in the anterior sacs, where arrival is delayed by one or two cycles. The usual composition of the air in each sac is given at the right-hand side of the graph. (From Schmidt-Nielsen *et al.* 1969.)

followed the arrival time of oxygen at various points with an oxygen electrode (figure 22). Usually some of the inhaled oxygen arrived at the posterior sacs towards the end of the first inhalation or very shortly thereafter. The oxygen concentration in the posterior sacs continued to rise through the expiratory phase of the first cycle, but there was no further increase during the second breathing cycle. This confirms that inhaled gas, which during the second cycle again was atmospheric air, arrives directly at the posterior sacs.

In the anterior sacs, however, arrival of oxygen was always delayed, it never occurred until some time during the second or third respiratory cycle. This means that the oxygen that finally appeared in the anterior sacs, must have been located elsewhere in the meantime. We concluded that air entering the anterior sacs therefore must come from the posterior sacs and/or the lungs, and summarized the results as follows: 'Inhaled air arrives first at the posterior sacs (although some may enter the lung directly); then, during exhalation, much of the air from the posterior sacs enters the lungs, and the anterior sacs receive air that has passed across the respiratory surface of the lung' (Schmidt-Nielsen *et al.* 1969).

The low respiratory frequency of the ostrich (about 5 cycles/min) permits another interesting observation, that the air-sacs are highly ventilated. With pure oxygen as a tracer gas, we found that both posterior and anterior air-sacs have a short washout time; in other words, none of the air-sacs contains an inert mass of stagnant air. A short washout time in the anterior sacs, which have a high CO_2 concentration and relatively low oxygen concentration (except during panting), must mean that they do not receive outside air. Their gas composition can best be explained by assuming that they are filled by air coming from the lung where it has passed across the gas exchange surfaces.

One of my collaborators in the ostrich study, Dr W. L. Bretz, then designed a probe to record the direction of air flow which was small enough to be placed at strategic points in the respiratory system of ducks. For further discussion I shall now return to the simplified diagram of the avian respiratory system which was used before (figure 20). Let me repeat that all anterior sacs are represented as a single space, and all posterior sacs as another. The main bronchus connects directly to the abdominal air-sac, and on its way gives off branches to the anterior sacs as well as to the lung itself. The air-sacs further connect directly to the lung through the recurrent bronchi, thus increasing the number of openings and connections.

The study of the air flow pattern in the bird lung indicates air movements as summarized in figure 23 (Bretz and Schmidt-

Nielsen, 1970, 1971). During inhalation both anterior and posterior sacs expand.* The main air stream flows to the posterior sacs which thus receive a mixture of dead-space air and fresh outside air. Part of the inhaled air also enters the lungs through the most posterior (caudo-dorsal) orifices. In the duck, as in the ostrich,

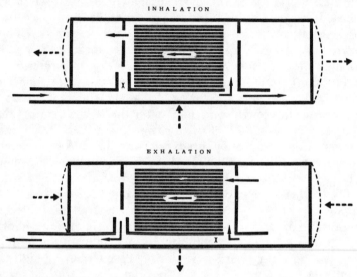

Figure 23. Air flow in the avian respiratory system. On inhalation (*top*) anterior and posterior air-sacs expand, outside air flows to the posterior sacs, and lung air flows into the anterior sacs. On exhalation (*bottom*) air from the posterior sacs flows into the lungs, and air from the anterior sacs is exhaled via the trachea.

The result of this flow pattern is that air flows continuously through the lung, both on inhalation and exhalation. The single broken arrow beneath each diagram indicates the movements of the pulmonary aponeurosis (see text).

there was no indication that outside air flowed to the anterior air-sacs during inhalation; therefore they would have to be filled by air from the lung. That no outside air entered the anterior sacs is consistent with the analysis of the gas composition in these sacs and with the fact that the flow probe placed in their connections to the main bronchus showed no flow (marked with an X in figure

* It has been suggested that the air-sacs could function as a set of bellows in alternating phases of contraction and expansion. For example, if the posterior sacs contract as the anterior expand, air would flow between them. However, there is a simultaneous and almost identical pressure drop in anterior and posterior sacs, and this precludes the direct flow of any major mass of air from one set of sacs to the other.

23). That there is air flow from the lung to the anterior sacs is an inference, however, consistent with the findings in the ostrich. Since the sacs do expand, air coming from the lung is the only remaining possibility. Also, the high CO_2 concentration in the anterior sacs (about 6%) is consistent with this conclusion.

During exhalation (figure 23) there is insignificant flow in the mesobronchus (marked with an X), and the posterior sacs must therefore empty into the lung itself. The anterior sacs, on the other hand, empty directly to the outside, as shown by the substantial flow in the appropriate bronchial connections.

If we propose that during exhalation air flows from the posterior sacs into the lung, we encounter one conceptual difficulty. The bird lung is said to be a relatively rigid structure which is supposed to undergo only minor volume changes during the respiratory cycle. It is somewhat difficult to see why the lungs have been dismissed so readily, for the thoracic region of birds seems to be built especially to permit large volume changes, a characteristic feature being the mobile, hinged ribs (figure 24). The common concept is

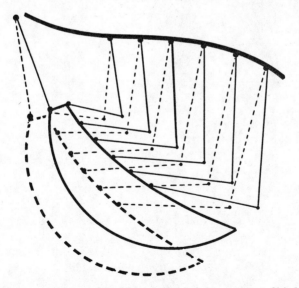

Figure 24. The arrangement of the hinged ribs and the sternum of birds permits large volume changes of the thorax, in which the lungs and some of the air-sacs are located.

that air movement in the lung is provided only by the bellows action of the air-sacs. It is consistent with this notion that birds

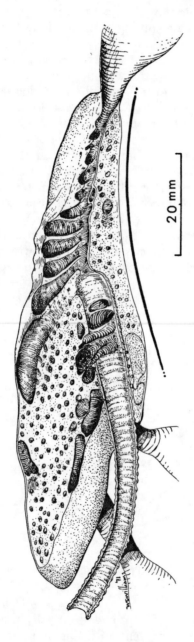

20 mm

Figure 25. Longitudinal section through a duck lung. Note the large secondary bronchi which connect with the posterior part of the mesobronchus. These secondary bronchi curve over the dorsal surface of the lung; they are quite thin-walled and will expand in response to a pull on the ventral surface of the lung by the aponeurosis (schematically indicated by the heavy black line).

have no muscular diaphragm; although a membranous structure, the aponeurosis, is located along the ventral surface of the lungs. This membrane arches slightly up into the lungs, and the lungs should expand on its contraction. Dr Bretz has recently suggested that muscles connected to the aponeurosis may be important in moving air from the posterior sacs into the lung during exhalation. Although the lung itself may expand little,* the large and numerous secondary bronchii on the dorsal surface of the lung could readily be expanded by muscles pulling on the aponeurosis (see figure 25). It is therefore particularly interesting that Soum (1896) observed contraction of the aponeurosis during exhalation, and that it recently was reported that the appropriate muscles are electrically active during exhalation (Fedde, Burger and Kitchell, 1963). We thus arrive at the apparent paradox that the bird lung is filled during exhalation and emptied during inhalation.

Control of air flow

Many authors have emphasized that the respiratory system of birds has no obvious valves that could help direct the air flow. For me, the conceptually most difficult detail is that, when pressure in the anterior sacs falls during inhalation, no air seems to flow from the main bronchus into these sacs. This conclusion seems certain, since the flow sensors show that there is no flow here, and inhaled tracer gases do not flow directly to the anterior sacs.

The entire system is a high velocity, low pressure system, and aerodynamic flow patterns therefore become important in directing flow. The orifices of the secondary bronchi seem eminently well designed from the view-point of fluidics, and their shape and direction may well be important in forcing the air stream during inhalation to remain attached to the main bronchus.

One aspect of importance is that the resistance to flow through the lung itself can probably change drastically. The inside walls of the parabronchi are covered with a lace-like network of smooth muscle fibers; contraction of these alters the diameter of the tubes. Mechanical stimulation causes contraction of the smooth muscle, and it can be made to contract or relax by various drugs (King and Cowie, 1969).

---

* The bird lung is more rigid than the mammalian lung; it retains its form and does not collapse on removal from the body.

In an attempt at evaluating the role of the parabronchi in changing air flow, we measured changes in their diameter in the duck lung. The muscles were relaxed by adrenaline infusion, and made to contract by infusion of methacholine (a cholinergic drug that persists longer in vertebrate blood than does acetylcholine). The measurements showed that the diameter of the parabronchi of the duck lung could change in a ratio of 1:2.3 (see figure 26).

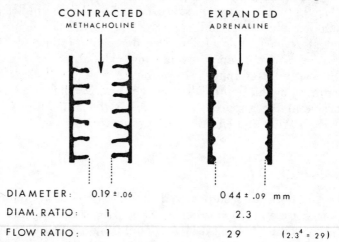

	CONTRACTED METHACHOLINE	EXPANDED ADRENALINE	
DIAMETER:	0.19 ± .06	0 44 ± .09 mm	
DIAM. RATIO:	1	2.3	
FLOW RATIO:	1	29	$(2.3^4 = 29)$

Figure 26. The inside diameter of the parabronchi in the duck lung is decreased by cholinergic stimulation (*left*) and increased by adrenergic stimulation (*right*). These changes result in a nearly 30-fold difference in resistance to air flow. (Measurements by D. E. Murrish.)

Since flow in a tube is related to the fourth power of the radius (Poiseuille's law), this will, at a given pressure, cause a nearly 30-fold change in flow. To what extent such changes aid in the regulation of air flow in the living bird remains to be studied.

The through-flow of air in the lung

The possibility of a unidirectional through-flow in the avian lung has been suggested many times before. It has even been suggested that the function of the air-sacs is to provide the lungs with a continuous flow of fresh outside air, thus supplying the largest possible amount of oxygen to the blood. This arrangement would by no means be desirable. The effect would be the same as vigorous over-ventilation of the lung, the blood would lose excessive amounts of CO_2, and acid–base balance would be grossly disturbed.

Normal respiration in birds and mammals virtually requires the blood to come into equilibrium with a reasonably high CO_2 concentration. This is precisely what would be achieved by the proposed system. The posterior sacs would serve as mixing chambers in which, on inhalation, air from the dead-space is mixed with additional outside air, thus obtaining a mixture containing some 3–4 % CO_2 (a concentration frequently found in these sacs by air analysis). This mixture, still high in oxygen, passes through the lung, permitting full oxygenation of the blood. On its exit from the lung, the air would be close to equilibrium with venous blood and have a high CO_2 concentration, about 6 % or even more. This air enters the anterior sacs, which thus serve as holding chambers from which the air is exhaled directly to the outside on the following exhalation.

Evidence for countercurrent flow?

The unidirectional flow of air in the bird lung raises the possibility of a countercurrent flow between air and blood. Such countercurrent flow occurs in the fish gill, where it has the advantage that arterial blood can become saturated with oxygen at the oxygen pressure of the inhaled, rather than the exhaled water. This is not possible in the mammalian lung, where the arterial oxygen can, at best, equal the oxygen pressure in the exhaled alveolar air.

In the lamellae of the fish gill, water and blood flow in opposite directions (figure 27). As a consequence, the blood, just as it is about to leave the gill, encounters the incoming water which still has all its oxygen; that is, the oxygen tension of the blood will approach that of the water before any oxygen has been removed. At the other end of the lamellae, the water that is about to exit encounters venous blood, so that, even though much oxygen has already been removed from the water, more can still be taken from it by the blood. As a result of this arrangement, fish may extract as much as 80 to 90 % of the oxygen in the water, an efficiency which could not easily be achieved without a countercurrent flow.

Evidence in favor of a countercurrent flow in the avian lung was obtained during our studies of the ostrich in which the arterial blood had a lower P_{CO_2} than we found in the anterior air-sacs. Many years ago, Zeuthen, who at that time was working in the same laboratory room as I in Professor Krogh's Institute in Copenhagen, observed that air which leaves the lungs during exhalation must be considerably richer in CO_2 than indicated by the CO_2 tension of the arterial blood (Zeuthen, 1942). To explain this he proposed a model for the blood flow which, in the end, results in the mixing of blood that has been in gas exchange with

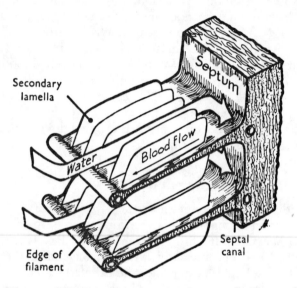

Figure 27. Water and blood in the shark gill flow in opposite directions. This is an example of a typical countercurrent flow which permits a higher degree of oxygenation of the blood than otherwise would be possible. (From Grigg, 1970.)

air of differing compositions. This system has recently been skillfully analyzed by Scheid and Piiper (1970) who referred to the principle as a 'cross-current' exchange system. The reason that Zeuthen did not arrive at the concept of a countercurrent system is probably that he believed in a reversal in the direction of air flow between inhalation and exhalation.

The constant and relatively low arterial P_{CO_2} in a number of bird species (close to 30 mm P_{CO_2} or less) contrasts with the higher P_{CO_2} (often above 40 mm P_{CO_2}) in the anterior air-sacs in a wide variety of bird species. Since the sacs are highly ventilated, the high CO_2 concentrations are not those of a stagnant, enclosed volume of air, they can only be the result of a gas-exchange pattern different from that of mammals. The conclusion can be extended to apply to birds in general by reference to a recent paper by Lasiewski and Calder (1971) in which they made a valuable contribution by compiling information on respiratory variables in a large variety of birds. Their generalizations are well analyzed and documented, and while they find that 'birds remove a greater proportion of the oxygen from tidal air than do mammals' they also note that the carbon dioxide tension in arterial bird blood is 'approximately 70% of that in mammals'. These generalizations are undoubtedly correct, but they cannot be achieved by a diffusion exchange of the mammalian type and a countercurrent flow seems necessary. However, the strongest support comes, I believe from the fact that birds can fly at altitudes of more than 6000 meters.

Bird flight at altitude

What advantages can birds derive from their complex respiratory system that are not achieved by mammals? We know that mammals and birds at rest have equally high metabolic rates (Lasiewski and Dawson, 1967). During flight the metabolic rates of birds is increased some 10-fold or more, but vigorous exercise in man causes similar increases in the oxygen consumption. We therefore cannot subscribe to any statement that birds need more oxygen than mammals and therefore must have a more 'efficient' respiratory system. Birds may have a higher ability for sustained high levels of activity, during long migration flights, for example, but this in itself should not require a different design of the respiratory system. An even better reason for concluding that the avian respiratory system is not a necessary prerequisite for flight is the fact that bats, with a perfectly normal mammalian-type lung, are excellent fliers and some species even migrate over long distances. The first determinations of the metabolic rate of bats in steady flight showed an oxygen consumption (27.5 ml O_2 g^{-1} h^{-1}) as high as in flying birds (Thomas and Suthers, 1970).

At high altitude, however, there is a conspicuous difference. A few years ago Dr Tucker at Duke University, while studying bird flight, exposed sparrows to a simulated altitude of 6100 meters (Tucker, 1968a). This represents 349 mm atmospheric pressure and a partial pressure of oxygen of 73 mm. At this altitude the sparrows showed no visible change in behavior, they were as alert and active as at sea level, and although they would not fly spontaneously, they were able to fly and gain in altitude if released gently into the air. In contrast, mice, that at sea level were con-tinuously active and exploratory, at 6100 meters were comatose, lay on their bellies, and could barely crawl.

The most obvious question to answer, in comparing these two animals, is whether the blood of the sparrows has a higher affinity for oxygen than that of mice, thus enabling a higher oxygen extraction from the air. This would be analogous to the llama of the Andes, in which the oxygen dissociation curve is displaced to the left in the customary plot, indicating a higher O_2 affinity which thus permits a higher degree of oxygenation of the blood at the low oxygen pressure of altitude. Tucker found, however, that the oxygen dissociation curve of sparrow blood is nearly identical to that of mouse blood, thus eliminating this possible explanation. On the assumption that the blood exchanges gases with the lung air in the same way as in the mammalian lung, Tucker calculated that the arterial blood of the sparrows at 6100 m altitude would be 24% saturated with oxygen. If, on the other hand, we recalculate Tucker's observations on the assumption that there is a counter-current flow between air and blood in the bird lung, the gas exchange would be different and the arterial blood of the sparrows would reach 80% saturation with oxygen (figure 28). This figure is

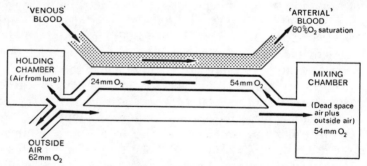

Figure 28. The flow of air and blood in opposite directions (countercurrent flow) in the bird lung permits the arterial blood to come into equilibrium with air of high oxygen content from the posterior air-sacs. This is particularly important for flight at altitude, as explained in the text. The figures on the drawing correspond to an altitude of 6100 m, corrected for 54 mm water vapor pressure (39.4 °C).

much more consistent with the oxygen saturation that we know is required for a mammal to remain conscious. Since the arterial blood may have contained three times as much oxygen as Tucker calculated, the estimated cardiac output and stroke volumes should be correspondingly adjusted downward.

In conclusion, then, we can say that both the avian and the mammalian respiratory systems have adequate capacities for extracting oxygen at sea level, even during vigorous exercise. At high altitude, however, the difference in the two systems becomes conspicuous. Flying birds probably require a 10-fold increase in metabolic rate, and yet birds can fly at altitudes where mammals are completely unable to maintain high activity and, in fact, are barely able to survive.

**The trumpeter swan, or
Dead-space and wing beat**

The mixing in the posterior sacs of dead-space air with fresh outside air has an interesting consequence – the size of the dead-space becomes physiologically rather unimportant. Then it is less puzzling than we might have thought at first glance, that good fliers such as swans and cranes have unbelievably elongated tracheas. In the trumpeter swan the trachea is in fact so long that it forms an S-shaped loop within the sternum (figure 29). Its total volume exceeds the value expected from the body size of this bird by 3.3 times (Hinds and Calder, 1971). Such an increase in dead-space for no obvious reason seems quite meaningless and invites some kind of rational explanation.

Let us see how it can be analyzed. A large dead-space requires a corresponding increase in tidal volume to achieve the desired mixture of dead-space air and fresh air on inhalation. For a given ventilation volume an increased tidal volume means a decrease in breathing frequency. Birds in general do in fact have much lower respiratory frequencies than mammals although their oxygen consumption is similar – for a body size of 1 kg the breathing frequency of birds is less than one-half that of a mammal of the same size (Lasiewski and Calder, 1971). In large birds the wing beat and respiration are probably synchronized (Tucker, 1968*b*). This is an old suggestion, derived from commonsense considerations of bird anatomy, but adequate information is not available for a large number of birds, and the following argument may therefore have to be modified. Nevertheless, for aerodynamic reasons the wing beat for a given bird remains rather constant even when the flying speed is changed. If respiration during flight is synchronized with the wings, it will be the wing beat that determines respiratory frequency, tidal volume must be adjusted to the low frequency, and in the end a large dead-space volume might then be required to achieve a suitable CO_2 concentration in the posterior air-sacs (the mixing chamber). We know that large birds, such as swans and

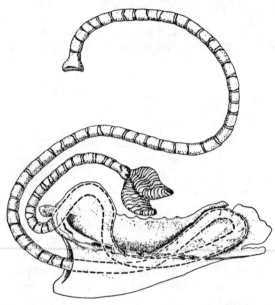

Figure 29. The trachea of the trumpeter swan is so long that it forms an S-loop within the sternum, thus greatly increasing the respiratory dead space. This may be no disadvantage to a good flier of large body size. (From Banko, 1960.)

cranes, have a very slow wing beat, and it is in these that we find the seemingly meaningless increase in dead-space represented by the inordinately long tracheas. Although this interpretation could be wrong, it seems no more unreasonable than saying that the long trachea aids in vocalization for many birds with more 'normal' tracheas are excellent vocalizers.

4. Exercise, energy, and evaporation

In this chapter I shall discuss the energy expenditure for animal activity, the effects of activity on the heat balance, and the need to use evaporation of water to keep cool.

Air-cooled flying machines

To conserve weight, man-made aircraft use air-cooled engines. No engineer would dream of loading an airplane with water to cool the engines by evaporation during long flights. Nevertheless, it has been said that birds produce so much heat during flight that they must use water to cool themselves. It has been calculated that a pigeon flying at 70 km h^{-1} requires about 8 grams of water for cooling for each gram of fat used as fuel, but this is based on untenable assumptions about the production and dissipation of heat in a flying bird. This becomes obvious if we consider long-distance migrations which may take place over many hundreds, perhaps even thousands, of kilometers, at times over open ocean without any opportunity to feed or drink. (For a recent, well-considered review of the energetics of bird migration, see Tucker, 1971.) When a migrating bird starts on its flight, as much as 50% of the body weight may be fat. If, in addition, it should need to carry several times the weight of the fuel as cooling water, the situation would be patently absurd.

There are quite a few reports that migrating birds seem severely dehydrated at the end of a long flight, but this claim seems based on wishful thinking and has never been adequately substantiated. On the contrary, investigators who have analyzed the carcasses of migrating birds which are killed by flying into television towers and other obstacles have reported that such birds by no means are particularly dehydrated although they often have flown for long distances and have exhausted their fat reserves (e.g. Odum, Rogers and Hicks, 1964). In fact, in a letter to me, Odum summed up many years of studies in the following words: 'Energy, not dehydration, is the limiting factor for the length of migratory flights. The water content as a per cent of fat-free weight remains near-constant in migratory birds, regardless of the stage of their migration'.

If we want to design an air-cooled bird, what steps can we take? It will be convenient to discuss this in general terms rather than specifically for flight. First of all, we will give birds a high body temperature, which facilitates the transfer of heat to the environment without evaporation. The normal body temperature of birds is some 40 to 42 °C and is sufficiently favorable to permit heat loss without evaporation, except under the most severe climatic conditions.

We can therefore, if we wish, regard the relatively high normal body temperature of birds as a factor which contributes towards a reduction in use of water for cooling; in other words, a reduction in the need for carrying extra weight during flight. The advantage is not limited to flight, however. Small rodents in hot climates escape the need for using water for cooling by escaping to an underground burrow. Few birds are burrowing, and a high body temperature greatly reduces the need to use water for cooling. (Many large birds, vultures, for example, escape the heat of the midday by soaring high in the air where the temperature is lower.)

Since water is inevitably lost by evaporation from the respiratory tract, another way to save water is to increase the oxygen extraction from the respired air. This would reduce the total ventilation volume needed at a given rate of oxygen consumption. As I suggested in the preceding chapter, the unidirectional flow of air through the bird lung permits a high oxygen extraction from the exhaled air, without a need to increase the oxygen affinity of the blood. It is quite possible that birds in flight can achieve an oxygen extraction high enough for evaporation from the respiratory tract to amount to less than the simultaneous formation of oxidation water; flight would thus be no extra drain on the water reserves of the migrating bird. At high altitude this might differ, for with lower oxygen pressure, ventilation volume would have to be increased to obtain the same amount of oxygen. On the other hand, the low air temperature at high altitude contributes to a reduction in respiratory evaporation, and to what extent this can counteract the expected increase we just do not know.

The cost of running

For many years heat regulation and its effect on evaporation of water has been studied in animals exposed to high temperatures in various kinds of heated boxes and chambers. The heat loads have been imposed from the outside, and increased internal heat load, such as that caused by exercise, has received relatively little attention, except for man and to some extent the dog. We therefore undertook a series of studies of animal activity and its effect on heat regulation.

To begin with, we determined the increase in oxygen consumption, and thus heat production, caused by running in a number of mammals ranging from mice to medium-sized dogs. As the animals ran faster, the oxygen consumption increased linearly with the speed of running with a different slope for each species (figure 30).

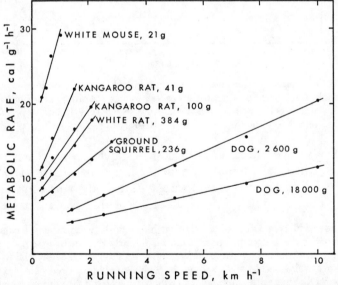

Figure 30. The oxygen consumption of a running animal increases linearly with the running speed. This permits us to calculate the net energy cost of running, which, as explained in the text, is constant for each species and decreases with increasing body size. (From Taylor, Schmidt-Nielsen and Raab, 1970.)

The increment in oxygen consumption for running a little faster is represented by the slope of the regression lines. The slope is the y-value over the x-value, or the oxygen consumption $(\text{ml } O_2 \text{ g}^{-1} \text{h}^{-1})$

divided by the speed (km h⁻¹). If we extrapolate the straight lines to zero running velocity, we get the y-intercepts, and by subtracting this value from the oxygen consumption at any given running speed, the excess can be regarded as the net cost of running at that speed. The cost of running a given distance, when calculated in this way, will be a constant, independent of running speed.* For the

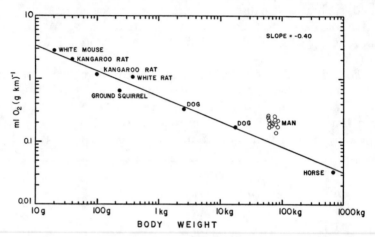

Figure 31. The net cost of running decreases regularly with increasing body size of mammals. Each point represents the cost of moving one gram of the body over 1 km (this is identical to the slopes of the lines in figure 30). The slope of the best-fitting regression line for the points is −0.40. (From Taylor *et al.* 1970.)

animals we have examined, this cost of moving is a greater deal higher for the small animal than for the large one (figure 31). The fact that a large animal can move 1 g of its body over 1 km more cheaply than the small animal should probably be considered as

* It is now widely acknowledged that within a fairly wide range of running speeds the oxygen consumption of man is linearly related to speed. At very high speeds the cost increases and the line curves upwards. Many claims of a non-linear relationship are based on the inclusion of high-speed points as well as a point for the resting metabolism (which differs from the y-intercept), thus showing a more complex, non-linear relationship. The fact that the y-intercept is higher than the resting metabolism can be ascribed to a 'postural effect' of maintaining the body in an upright position, for example, the metabolic rate of a standing man falls close to the y-intercept of the running curve. The question of 'true' resting metabolism is in fact one of great difficulty, in particular for animals whom we cannot tell to lie down and relax. Sleep and anesthesia are of little help, both give lower values than the resting, or very ineptly named 'basal' metabolic rate.

one of the evolutionary advantages of a large body size, an advantage which, as far as I know, has mostly been overlooked.

The size range of our animals was from mouse to dog, but after calculating the best-fitting straight regression line for these, we found that data for sheep and horse already in the literature fell remarkably close to an extension of our line. Curiously, available data for man fell uniformly at a level of about twice what we would expect for a four-footed animal of his weight. If man expends twice the expected energy for moving 1 g of his body weight over 1 km, what is the reason? Is this deviation due to a difference between bipedal and quadrupedal locomotion? If this is so, is the extra energetic cost of bipedal locomotion, the price man has paid for freeing one set of his appendages for other purposes, a step which, however, has given him nearly infinite advantages over those mammals that have remained quadrupedal? The question could be examined, but perhaps not settled, by reference to running birds which also run on two feet. So far we have only studied one running bird, the South American rhea, which weighs between 20 and 30 kg. This bipedal runner also turned out to move at higher cost than predicted from our mammalian regression line. It will be interesting to see whether this will hold for other birds as well.

Total or net cost

I should undoubtedly add that the preceding was based on an analysis of the increment in energy cost caused by locomotion; but an animal that moves, a grazing sheep for example, must continuously support its total metabolic rate. This would be analagous to common practice for automobiles; we measure the amount of fuel required for a certain number of kilometers, without subtracting the cost of idling the engine at zero speed. The choice between calculating the increment cost of locomotion, or the total use of energy while moving, depends on one's needs. Tucker, for example, recently used the total fuel cost to compare the cost of transportation for a variety of animals, using flying insects and birds, walking and running mammals, and for comparison, some man-made vehicles as well, just as Weis-Fogh did years ago when comparing insects to aircraft (Weis-Fogh, 1952). Tucker's elegant studies of flying birds contributed important information which greatly extended our knowledge of animal locomotion (Tucker, 1968b). Data previously compiled by Tucker (1970) have been combined with other information in figure 32, thus giving an overall view of the cost of animal locomotion.

There is one aspect of the grouping of the points for mammals that needs further discussion. It seems that the values for the smallest as well as the largest mammals are located well above a straight

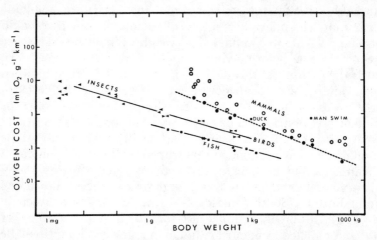

Figure 32. The cost of animal locomotion, expressed as the amount of oxygen needed to transport 1 g animal over a distance of 1 km. Except where noted, these data refer to total metabolic expenditure while moving.
Circles – walking and running. Black circles refer to the net cost of locomotion (same data as in figure 31), regression line calculated by method of least squares. Open circles refer to the total oxygen consumption of animals while moving and thus include the resting (or zero speed) O_2 consumption. These points therefore fall above the regression line for net cost. (Calculations based on same sources as Tucker, 1970; additional points added from other sources.)
Arrowheads – flying animals. Double arrowheads, birds; single arrowheads, insects. Regression lines fitted by eye.
Squares – swimming animals (fish, except as identified for duck and man). The five squares along regression line (eye-fitted) are for salmon (Brett, 1965). The two diamonds refer to whitefish and greyling (Matyukhin and Stolbow, 1970). The twin-diamond represents pinfish (*Lagodon rhomboides*, a non-salmonid fish) (Wohlschlag *et al.* 1968). Square marked 'duck' from Prange and Schmidt-Nielsen (1970).

line, the entire distribution therefore being curved upwards at both ends. The black circles in figure 32 are the same as those used in the preceding figure (figure 31) with a few more points added from available publications, and the calculated regression line for these points refers to the net cost of locomotion as I discussed it above. For the calculation of the open circles the total oxygen consumption of the animal was used, and the resting (or zero speed) metabolic rate is therefore included in the cost. If the non-moving rate were subtracted, these points would be displaced downwards towards the regression line.

Here we have an explanation for the seemingly curved distribution of the points. The points about the center of the size range are from animals that ran at rates of up to 6- or 8-fold increases in metabolic rate; the non-moving metabolism therefore is only a small fraction of the total, and it makes only a minor difference whether or not this value is subtracted. At the lower end of the size range, many of the animals could not be made to run at more than from 2 to 3 times the non-moving metabolic rate, which therefore makes up a substantial fraction of the total. Had this value been substracted, the points would be displaced downwards to a considerable extent. At the other end of the size range the values are for walking horses. Again, the non-moving metabolic rate makes up a sizeable fraction of the total during walking, giving a strong upward bias to the points. Subtraction of the non-moving value would bring them down towards the regression line for the net cost. It is therefore probable that the curved distribution of the points for mammals is an artefact, due to the inclusion of the non-moving rate in the total cost.

It is not possible to say whether one way of calculating the cost of transportation is more correct than another. If we want to know how much fuel an animal consumes while walking, the total metabolic rate will be the correct measure. If, on the other hand, we want to know the energetic cost of locomotion as such, it seems more appropriate to examine the increment in metabolic rate for an increment in speed. Over a wide range this increment remains constant (represented by the straight lines in figure 30), and the cost of moving as opposed to not moving is easily calculated as the net cost of locomotion.

Flying and swimming

Let us now consider other modes of locomotion. It is not intuitively obvious why a flying animal can move more economically than a mammal of the same size, for a bird must continuously support his full weight against a fluid medium of low density. Merely to remain suspended in air costs a continuous output of power, and the aerodynamic drag of moving through the medium must also be overcome; for a running mammal the air resistance is a barely noticeable small fraction of the cost of moving at normal speeds. If, however, we consider that a migrating bird probably can fly more than 1000 km non-stop, it is easy to grasp the economical advantage of flight, we could not possibly imagine a mouse of the same body weight running 1000 km without stopping or feeding.

While the cost of running increases regularly and linearly with increasing speed, the situation is different for birds. Budgerigars have an optimum speed of flight at which their oxygen consumption is at a minimum. Flying slower or faster increases the metabolic rate (figure 33). Although flying level at 35 km h^{-1} gives the lowest

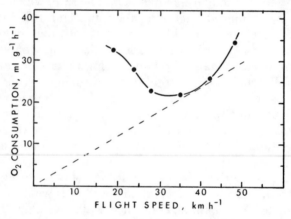

Figure 33. The oxygen consumption of a flying budgerigar is at a minimum at 35 km h^{-1}; flying faster or more slowly increases the oxygen consumption. (The most economical speed, the speed at which the bird can cover the greatest distance at the lowest cost, is at a somewhat higher flying speed, as explained in the text.) (Based on Tucker, 1968*b*.)

metabolic rate, the speed that enables the bird to cover a certain distance most economically will be higher, in the case of the budgerigar about 40 km h^{-1}. This should be no surprise; after a moment's thought we realize that a bird flying faster gets there sooner. In fact, the optimum speed to cover a certain distance most economically can be found by simply drawing a tangent from the *x–y* origin to the curve in figure 33; the point of contact gives the lowest slope of this line, i.e. the lowest cost of transport.

A surprising aspect of flying is that the points for insects seem to fall entirely on an extension of the regression line for birds (see figure 32). A few of the points for the smallest insects, those of about

1 mg size, fall below the line, but all these points are from rather old determinations. In contrast, the most recent data for flying mosquitoes, which weigh about 1 mg, fit the regression line exactly (Nayar and Van Handel, 1971). Since there are substantial differences in the aerodynamics of bird and insect flight, the ease with which we can draw a common regression line for them may be fortuitous rather than an expression of a fundamental similarity.

Swimming turns out to be far less expensive again than flying (see figure 32). Most swimming organisms have a specific gravity near that of water, and when they propel themselves the greatest output of power is needed to overcome the drag, and little, if any, goes into keeping from sinking. As drag increases with the square of velocity, the power requirement increases rapidly with the speed of swimming. Drag is also a function of surface area (as well as other factors such as shape), and larger fish should therefore, owing to their smaller relative surface, propel themselves more easily and with less effort.

Determinations of the cost of swimming for salmon ranging in size from a few grams upwards, do indeed show that it is more expensive to be a small fish (Brett, 1965). Other salmonid fishes (whitefish and greyling), examined by other investigators in a different part of the world, give results very similar to those for salmon (Matyukhin and Stolbow, 1970); this gives us considerable confidence in the investigators as well as the reliability of their determinations. Furthermore, a non-salmonid, the pinfish, also seems to adhere to the same pattern (Wohlschlag, Cameron and Cech, 1968). This makes it seem that various fish know perfectly well what it is supposed to cost to swim.

If we compare the great uniformity of the cost of swimming for fish, with that for two surface swimmers, man and duck, we find the latter far more costly. The swimming cost for a duck places it in the range expected for a walking animal – nearly 10-fold as high as a swimming fish. Man, with his body basically unfit for aquatic locomotion, finds swimming at least 5-fold as expensive as running, which goes to show that swimming is mostly for fishes as flying is for the birds.

I have taken pleasure in calculating, on the same basis as for fish, the cost of swimming for sperm. Using available information about the mechanical work done by sperm in propelling itself and information about its oxygen consumption (Rikmenspoel, Sinton and Janick, 1969; Rothschild, 1953), it is easy to calculate the

energy used by 1 g sperm in travelling 1 km. This calculation is
based on the total oxygen consumption of the sperm, assuming that
most of it goes into propulsion and that the sperm does little but
swim – anyway, that is most of its purpose in life. Interestingly,
the calculated point falls exactly on the extension of a line covering
the cost of swimming for salmon (figure 34).

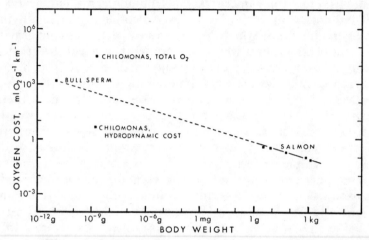

Figure 34. The cost of swimming for sperm falls exactly on an extension of the
regression line for swimming fish. For an organism in which only a small fraction
of the energy consumption goes into swimming (in this case the flagellate
Chilomonas), use of the total oxygen consumption places the point too high, use
of the calculated energy requirement places the point too low.

For an organism in which only a minor and unknown fraction
of the energy goes into swimming, we are in difficulties. The flagellate *Chilomonas* probably uses only a small part of its energy for
propulsion, and if we use the total oxygen consumption we get a
point far above the expected value: if we use only the amount of
energy that from fluid-dynamic considerations goes into propulsion
of this organism at its normal speed, the point is as far off in the
other direction (calculations based on Hutchens, Podolsky and
Morales, 1948).

Activity, body temperature, and evaporation

To return to land animals, I shall now discuss the effect of
activity on the use of water for evaporation by asking the question:
Does the increased respiration during running influence evapora-

tion? In the introduction to the first chapter I mentioned that for the kangaroo rat the most important avenue for water loss is the respiratory tract. Increased activity increases respiration and therefore evaporation as well, but the extent of increase is not easily predictable. Respiratory evaporation depends on the temperature of the exhaled air, which in turn depends on heat exchange in the nose. An increased rate of air flow through the nose could decrease the effectiveness of heat exchange, increase the temperature of the exhaled air, and thus increase the water content of the exhaled air.

At rest the kangaroo rat is in positive water balance because oxidation water is produced in excess of evaporation from the respiratory tract; the problem is whether, during activity, increased evaporation exceeds increased formation of metabolic water. Oxygen consumption can be used as a measure of metabolic water formation, and the ratio of evaporation (mg H_2O) to oxygen consumption (ml O_2) is therefore a convenient measure of water balance. If this ratio is less than about 0.6 or 0.7, kangaroo rats are in positive water balance; if it is higher, they are in negative water balance. (The exact break-even point depends on the composition of the food and its moisture content, for even dry seeds contain some water.)

When kangaroo rats were running at moderate temperature, the $H_2O:O_2$ ratio did not change appreciably with running (figure 35). In other words, if the animal at rest is in water balance, increased respiration during running does not alter the water balance. Increasing the air temperatures, however, makes the $H_2O:O_2$ ratio increase, and the water balance becomes increasingly precarious. However, since the ratio at any given temperature remains unchanged with running, activity as such has little direct effect. Should the evaporation increase beyond the amount formed in oxidation processes, however, continued activity increasingly deprives the animal of water, and it would do better by stopping and retreating to its underground burrow. I do not propose that kangaroo rats are nocturnal because of their water balance problems, but their water balance is certainly more favorable because their activity and search for food is restricted to the relatively cool night hours.

When it comes to animals that are active during the day and lack burrows in which to hide, evaporation becomes much more critical. To study this question we selected two medium-sized animals for laboratory use, the dog-like African carnivore, *Lycaon pictus*, commonly known as the Tricolored Cape Hunting Dog (although zoologically it is not a true dog), and a large bird, the ostrich-like South American rhea. Both are good runners and could be trained to run on a treadmill.

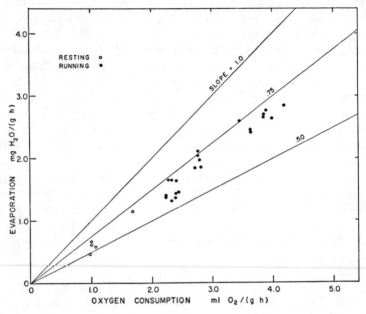

Figure 35. The evaporation from a kangaroo rat which runs at various speeds increases in the same proportion as the oxygen consumption. This means that the ratio of water loss to oxygen consumption remains constant, and an animal which is in positive water balance at rest will remain in positive water balance while running. Measurements at 25 °C. (Based on Raab and Schmidt-Nielsen, 1971.)

Years ago we found that the camel reduces its water loss during a hot desert day by permitting its body temperature to rise instead of expending water to keep cool (Schmidt-Nielsen *et al.* 1957). Of course, the camel can permit only a limited rise in body temperature; the maximum we observed was an increase of 7 °C, from 34 °C in the morning to 41 °C in the hot afternoon. A high body temperature has several water-saving advantages. First of all, no water is used to dissipate heat that goes into raising the body temperature, and the excess heat can be lost again without use of water during the cooler night. More importantly, with a high body temperature the gradient for heat flow into the animal from a very hot environment has been reduced and may even be reversed, thus permitting heat loss without use of water. Can an animal that produces a great deal of heat internally due to running also gain an advantage from a high body temperature?

The Cape Hunting Dog is a notorious predator that hunts during the heat of the day. It relies on endurance rather than speed as it tirelessly runs down its quarry. To maintain a constant low body temperature while running across the hot African plains would require high rates of evaporation, and this would presumably greatly limit the distance over which a hunting dog can pursue its prey. A high body temperature during running, however, would give the hunting dog an obvious advantage in its water economy.

The hunting dog is a rather intractable animal which has never been successfully domesticated. Captive mothers invariably kill their offspring, and only completely hand-reared animals have been raised in captivity. We acquired a newborn pup from the National Zoo in Washington, and rapidly discovered why raising hunting dogs is not a popular pastime, the odor alone prevents domestication. At some olfactory inconvenience to our surroundings our hand-reared animal was trained to run on a treadmill while wearing a mask so that its oxygen consumption could be determined. An account of its heat balance showed that the hunting dog indeed achieves considerable savings in use of water by permitting its body temperature to increase (Taylor *et al.* 1971*b*). When resting at room temperature the hunting dog had a body temperature about a degree higher than domestic dogs, and with increasing running speed the temperature of the animal increased faster (figure 36). At 15 km h^{-1} the temperature of an ordinary dog had increased by 1.3 °C, but that of the hunting dog had increased by 2.2 °C, reaching 41.2 °C. These two animals were running at room temperature; what difference was there in their heat balance?

The heat production (determined as oxygen consumption) of the running hunting dog was somewhat higher than that of the domestic dog, but when we look at the use of water (figure 36, lower part), we find that in the hunting dog running at 15 km h^{-1} only a quarter of the heat production was dissipated by respiratory evaporation (panting), while in the domestic dog well over a half of the heat production (66%) was dissipated in this way. The conclusion is obvious, the hunting dog does indeed reduce its use of water by permitting the body temperature to increase, thus facilitating heat loss by conduction. If water loss is reduced by one-half, the hunting range is presumably doubled (assuming that water is the limiting factor in the distance over which a hunting dog can pursue its prey).

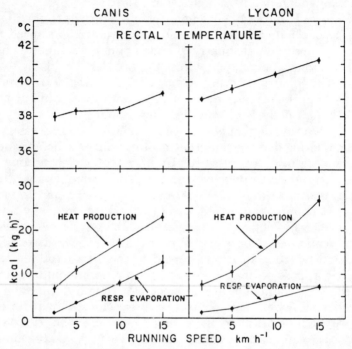

Figure 36. During running the Cape Hunting-Dog (*Lycaon*) reaches higher steady state body temperatures than the domestic dog (*Canis*) at the same speeds (*top graphs*). This is reflected in a relatively lower respiratory evaporation in the hunting dog (*bottom graphs*). Experiments at 26 °C, 25% rel. hum. (From Taylor *et al.* 1971*b*.)

On previous occasions I have often emphasized that an increased body temperature during heat stress cannot be considered a measure of failing temperature regulation (Schmidt-Nielsen, 1964). We now see that in running animals as well, an increase in body temperature can readily be interpreted as a means for reducing the loss of water, rather than as an expression of inadequate heat dissipation.

It has been known for many years that the body temperature of man during exercise is increased, and that the increase is closely related to the intensity of exercise. When steady state has been attained, the rectal temperature is accurately maintained at the higher level, the exact level being determined by the intensity of exercise (Nielsen and Nielsen, 1962). The usual interpretation is that the increased body temperature speeds up all physiological processes and therefore gives an advantage in performance. I should like to propose that, for an organism that has evolved as a hunter pursuing its prey by running it down in open country, the advantages of increased body temperature to the water economy may be a consideration of no less importance.

A running bird

Let us next consider the rhea, which is fairly large, some 25 kg, but not as large as the African ostrich which may reach 100 kg. In nature the rhea is a very fast runner, but it differs from man and dog in being a sprinter rather than a long-distance runner. It is therefore of interest to examine what means this animal has adopted for temperature regulation.

We found it difficult to train rheas to keep running persistently and at even speed, and they were unwilling to run for more than short periods at high temperatures. We therefore adopted an experimental strategy of running for 20 minutes, followed by 40-min recovery, making a total of one hour during which metabolic heat production and heat dissipation were followed.

Running for 20 min always resulted in a considerable increase in the body temperature of the rhea, at room temperature.the increase was about 2 °C, and at 43 °C slightly more, 2.5 °C (figure 37).

Since the rhea, when it runs at an air temperature of 43 °C, ends up with a body temperature slightly higher than the surroundings, some body heat can flow to the surroundings by conduction and radiation, and this reduces the amount of heat to be dissipated by evaporation. For a bird resting at 43 °C on the other hand, the body temperature remains below the ambient, heat flows by conduction into the animal and this extra heat, in addition to the metabolic heat, must be dissipated by evaporation. Let us now examine what effect the changes in body temperature will have on the use of water.

Running at 5 km h⁻¹, a moderate speed for a rhea, caused a 2½-fold increase in the metabolic heat production, irrespective of ambient air temperature. During one hour of 20-min running followed by 40-min recovery, the evaporation from the rhea increased with the air temperature in regular fashion (figure 38, right). In the resting bird, evaporation also increased with air temperature, but not as much (figure 38, left). If we compare the heat production and the amount of heat lost by non-evaporative means (conduction and convection), as indicated by the cross-hatched areas in figure 38, this amount is much greater for the active animal, primarily due to the increase in body temperature.

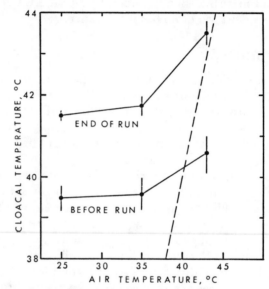

Figure 37. When the ostrich-like rhea runs, its body temperature always increases considerably. Running at 5 km h⁻¹ for 20 min caused the body temperature to increase by 2 °C or more. (From Taylor *et al.* 1971*a*.)

The amount is nearly three times as great in the bird after running, as compared to the resting bird. The area shaded with small dots represents heat gain from the environment. As I explained above, the resting bird, owing to its lower body temperature, has a heat gain from the environment at 43 °C air temperature, while the bird after running, owing to its higher body temperature, avoids this extra heat load.

The birds running at high air temperatures did not reach a steady state. They were unable or unwilling to run much longer, their body temperature did not stabilize, and they would probably be unable to run for prolonged periods in very hot surroundings. To what extent the rhea would become overheated if forced to run longer in the heat and how high body temperatures could be tolerated, are questions which have not yet been answered. It is possible that very high body temperatures could be tolerated if the birds were able to keep their heads cool, as I shall discuss next.

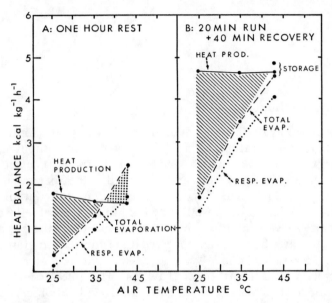

Figure 38. Heat balance in the rhea. The difference between heat production (*solid line*) and heat loss due to evaporation (*dashed line*) equals heat transfer by conduction and radiation (*cross-hatched area*). In the bird at rest (*left*) there is a heat gain from the environment (*dotted area*). The bird after running (*right*) has an increased body temperature, and the temperature gradient between body and the environment now permits the bird to continue losing heat by non-evaporative means as indicated by the cross-hatched area. (From Taylor *et al.* 1971*a.*)

5. Countercurrent, a cheap trick

Keeping the head cool

Recently Taylor has shown that gazelles and antelopes may reach body temperatures over 46 °C, apparently without harm. The oryx (*Oryx beisa*), which weighs about 100 kg, is extremely tolerant to hot dry conditions and can stand in the hot sun all day. When kept on a regimen of dry fodder without drinking, its oxygen consumption decreases substantially, and this reduces the internal heat load. At very high air temperature, however, the oryx will evaporate water by panting, but since it permits its body temperature to increase to above 45 °C, the amount of water used is drastically reduced. At times Taylor's oryx tolerated a rectal temperature of 46.5 °C for as long as 6 hours without observable ill effects.

How can a mammal withstand such high internal temperatures? The brain and central nervous system are probably the parts most sensitive to heat, and a tolerance of 46 °C or more would be unique among higher animals (Taylor, 1970). The answer is simply that the brain does not reach the same high temperature as the body core. In a gazelle in which the body temperature had been driven up by exercise, the brain temperature proved to be as much as 2.9 °C lower than in the central arteries (Taylor, 1969). The blood running to the brain is cooled owing to a peculiar arrangement of the blood vessels in the head. The external carotid artery, which supplies most of the blood to the brain, passes through the cavernous sinus, here it divides into hundreds of small arteries which again join into one vessel before entering the brain. The cavernous sinus is filled with venous blood that drains from the nasal passages, where it has been cooled by evaporation from the moist mucous membranes. It is inevitable that the arterial blood that runs through this pool of cool blood thus is cooled on its way to the brain.

The role of the carotid rete in achieving a difference between core temperature and that of the blood supplying the brain has been demonstrated in several animals by Baker and Hayward (1968). By a series of careful measurements on sheep, in which the circulation is similar to that in other ungulates, they showed that venous blood returning from the nasal mucosa to the cavernous sinus indeed cools the arterial blood in the carotid rete, and that this is important in maintaining hypothalamic temperature in the panting sheep at a tolerable level. Similar rete structures, which can be cooled by venous blood draining from the nasal region are, in fact, common to a large number of animals, but are not found in all (see figure 39). The dog has a more complex circulation, with

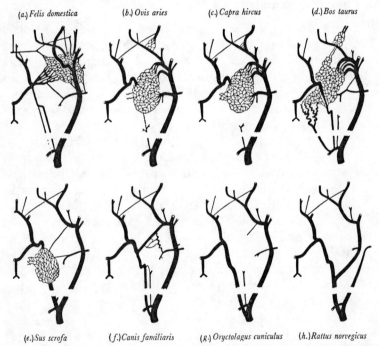

Figure 39. In many animals the arterial blood supply to the brain forms a network, or rete, which facilitates heat loss from the arterial blood. Blood therefore can enter the brain at a relatively low temperature and thus protect this sensitive organ from over-heating. (From Daniel, Dawes and Prichard, 1953.)

anastomoses between the internal and external carotids, permitting what Baker and Hayward described as 'two types of brain temperature regulation'. The rat has no rete or rete-like structure, but the device would be of little use to a non-panting animal which, as I discussed in the first chapter, has a nasal passage more suitable for conserving heat than for losing it.

It seems that the heat exchange in the carotid rete endows the panting animal with a major advantage over an animal that cools itself by sweating. The heat exchanger permits maintenance of a low temperature in one part of the body, in this case the brain, and this obviously reduces the total amount of water used. The sweating animal, it seems, must cool its entire body to the temperature it wishes to maintain in its brain, and thus must use larger quantities of water.*

More heat exchangers

The heat exchanger of the carotid rete is in principle analogous to the heat exchanger in the flipper of a whale, which was mentioned on page 6. In both cases a part of the body, in one case the brain, in the other the flipper, is kept at a temperature below the main bulk of the body. The difference is that while the whale keeps the flipper cool in order to conserve heat for the body core, the antelope keeps the head cool in order to permit the core temperature to rise more than could otherwise be tolerated.

It is also analogous to the heat exchange in the vascular bundles which supply the testes of sheep and are particularly well developed in many large marsupials (figure 40). A relatively small evaporation from the surface of the scrotum can keep this part of the body below core temperature, and the vascular heat exchanger prevents this 'cold' from being lost into the bulk of the body. Evaporation from the scrotal surface would cool the body core only a very little, and, in the absence of a heat exchanger, the testes would be supplied with blood at core temperature, which for poorly understood reasons prevents production of normally fertile sperm.

* The panting animal enjoys another advantage which I have discussed on earlier occasions, it can maintain a high skin temperature and this automatically reduces the gradient from the hot environment to the skin, thus reducing the environmental heat load and the use of water. The sweating animal, on the other hand, maintains a relatively low skin temperature (below body-core temperature), and this facilitates the flow of heat from a hot environment to the skin, thus increasing the heat load and use of water. However, we still lack a clear understanding of the relative biological merits of sweating and panting. It seems that sweating must have some advantage. Could it be that it aids in keeping the skin of a very thinly furred or naked animal from becoming over-heated in the sun?

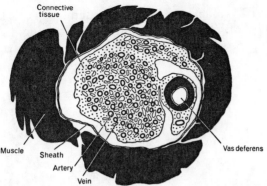

Figure 40. The vascular supply to the testis of large marsupials, in this case a wallaby, is through rete structures in which heat exchange can take place and, in fact, is inevitable. The most reasonable explanation of this structure is that it helps keep the testes at a temperature below that of the body core, which is necessary for the formation of fertile sperm. (From Barnett, Harrison and Tomlinson, 1958.)

Warm-blooded fish. Independent control of temperature in limited parts of the body is also responsible for the high muscle temperatures in some large fast-swimming fish, such as tuna and sharks. Most fish have a body temperature very close to that of the surrounding water. The blood that takes up oxygen must flow through the gills where it is cooled to water temperature, and it is impossible for a fish to achieve a high body temperature unless a heat exchanger is imposed between gill and tissues. In the tuna, the blood vessels that supply the dark lateral muscles* form a sheet of fine blood vessels in which arteries are densely interspersed between veins, with the blood running in opposite directions in the two kinds of vessels (figure 41).

This heat exchanger is supplied with arterial blood coming from the gills and running in vessels close to the surface of the fish, and thus containing blood at water temperature. The cold end of the heat exchanger is therefore on the outside of the fish, and the warm end is in the interior muscle mass. As a result, the tuna can maintain muscle temperatures as much as 14 °C above the water in which it swims (Carey and Teal, 1966). A similar system is effective in the mako shark (*Isurus oxyrhynchus*), and by inference in several other

* Rayner and Keenan (1967) reported that the basal swimming of the tuna is carried out entirely by the dark muscle and that the white muscle represents a reserve of power for short bursts of high activity.

species of shark which have vascular structures of this kind (Carey and Teal, 1969). These large and fast-swimming fish therefore probably enjoy some of the advantages of being 'warm-blooded'.

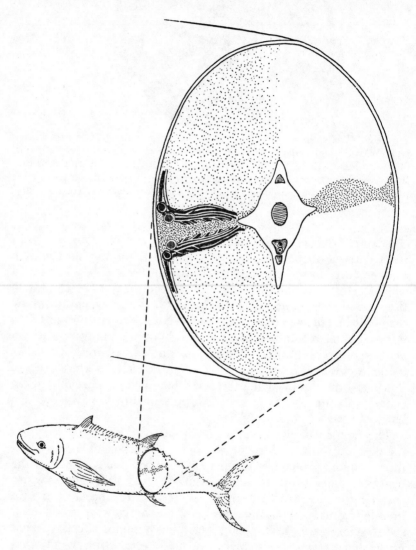

Figure 41. The blood supply to the red muscles of the tuna is through a heat exchanger which permits these muscles to remain warm. Fine arteries which originate from a larger cutaneous artery run inwards in a sheet in which the arteries are densely interspersed with veins. (From Carey and Teal, 1966.)

Oxygen – a problem

One question has so far not been raised. If, for example, we look at the flipper of the whale, heat readily diffuses from arterial to venous blood, thus never reaching the periphery. If oxygen were to diffuse equally well, an effective heat exchanger would not permit oxygen to reach the peripheral tissues. In fact, in the kidney the easy diffusion of oxygen between the blood capillaries supplying the renal papilla and those leaving it (descending and ascending vasa recta) must be responsible for the very low oxygen found in this tissue, in spite of its large blood supply (Landes, Leonhardt and Duruman, 1964).

As we examine heat exchangers, however, we find that they consist of blood vessels with a diameter of about 1 mm, more or less. We know that oxygen does not diffuse readily through the walls of vessels of this size, and a heat exchanger with vessels of these dimensions, therefore, is not a good exchanger for oxygen. This can be expressed in quantitative terms by comparing the diffusion coefficients for heat and oxygen, provided we can establish a reasonable equivalence between the two as blood flows through the exchanger.

The oxygen capacity of mammalian blood is about 200 ml per liter of blood. The oxygen in one liter of blood, if fully utilized (the physiologically possible maximum) will produce 1000 calories, in other words, it could raise the temperature of the one liter of blood by one degree, but no more. We now have established a kind of physiological equivalence for oxygen and temperature. A gradient of oxygen of 150 mm (the difference between fully arterialized blood and zero pressure in the tissues) can produce a temperature gradient of 1 °C, for, in the steady state, each liter of blood can carry to the tissues the above-mentioned oxygen and carry away the equal amount of heat. If the oxygen capacity of the blood is lower (in fish, for example, it is about half that of mammals), the maximal temperature gradient will be reduced accordingly, but in principle the conditions will be the same.

We are now ready to examine whether the diffusion of oxygen in a heat exchanger is so great that a whale flipper, or even worse, the testes of a kangaroo, gets little or no oxygen. If a heat exchanger has an efficiency of 99%, it means that 99% of the heat arriving in the arterial blood is returned to the venous blood and only 1% passes to the peripheral end of the exchanger. A simple calculation will show that an exchanger which permits the transfer of 99% of the heat potentially available from the oxygen in the arterial blood, will permit only 2% of the oxygen carried by the arterial blood to diffuse to the venous blood and thus be kept from arriving at the peripheral end of the exchanger and the distal tissues.

An efficiency of 99% in a heat exchanger is very good indeed; the loss of 1% of the heat carried in the arterial blood to a whale flipper, for example, is quite insignificant to the animal. The fact that 2% of the oxygen of the arterial blood is returned to the veins, thus not reaching the periphery, is likewise insignificant. If the efficiency of the heat exchanger were less, the amount of oxygen 'lost' from the arterial blood would also be less, for the amounts of heat and oxygen must change in the same proportion for the simple reason that the diffusion coefficients remain the same. Thus, a heat exchanger that is only 95% efficient, letting 5% of the arterial heat pass through the exchanger, will permit no more than 0.4% of the oxygen of the arterial blood to diffuse across to the venous blood.

If we now turn to those countercurrent mechanisms that involve the diffusion of solutes rather than heat, we find their dimensions to be of capillary size, say, one-hundredth of a millimeter or so. The smaller size of the vessels has a 2-fold effect, the total wall area is increased because of the much larger number of the vessels, and the wall through which diffusion takes place is much thinner. In the rete mirabile of the fish swimbladder, which I shall soon discuss in detail, the diffusion of gas is excellent between the thousands of densely packed thin-walled capillaries. In the bird lung the diffusion of gases between blood and air takes place across a distance of about 1 μm, no major barrier to the rapid diffusion of gas. In the nose of small mammals, heat and water vapor move rapidly; the passageways are larger than capillary-size, but the diffusion of water vapor in air is some 100 000 times faster than the diffusion of a gas through tissues.

In summary then, countercurrent systems used for heat exchange utilize vascular channels of the millimeter size range, which means that they cannot function as gas exchangers. The relatively slow diffusion of gases in a heat exchanger is an advantage; otherwise they could not be as widely utilized as they are. For gases, efficient exchangers must be at the capillary level.

Exchangers and multipliers – a cause of confusion

All the heat and gas exchangers we have examined so far depend on 'down-hill' flow, from a warmer to a cooler place, from a higher to a lower gas concentration. It has been the cause of a great deal of confusion that certain structures that can work 'up-hill', i.e. move a substance against its concentration gradient, have structural similarities to exchangers and indeed also have some of the same properties.

These structures, which I shall now discuss, are known as multipliers or, to be more precise, countercurrent multipliers. In the mammalian kidney, for example, there is an active transport of sodium ions that ultimately is responsible for the formation of a concentrated urine; the anatomical arrangement is such that a modest transport is gradually built up to a high concentration, hence the name multiplier. In the swimbladder of fish, gas can be secreted against pressures of several hundred atmospheres, again with the aid of a multiplier system. It is worth the effort to make it quite clear why and how these multipliers differ from an exchange system; the similarities will then be readily apparent and should cause no further confusion.

The kidney. Lower vertebrates are unable to form a urine which is osmotically more concentrated than the blood plasma, only birds and mammals can do this. The mechanism is described in most modern physiology texts, and there is no need to explain the details here, a few words about the overall picture will suffice.

In the mammalian kidney each nephron forms a hairpin loop, Henle's loop. The hairpin has two legs, one leading into the loop and called the descending leg because it descends into the renal papilla; the other, the ascending leg, leads out of the loop. Sodium ion is removed from the fluid in the ascending leg by active transport, called 'active' because it is from a lower to a higher concentration and requires energy. This increases the sodium concentration in the surrounding tissues, and sodium can now enter the descending leg by simple diffusion. The fluid leaving the loop has had sodium removed from it continuously, and thus contains less sodium than the fluid which enters the loop. Sodium must therefore accumulate in the loop itself in a higher and higher concentration. In this way a very high concentration can be built up in the papilla

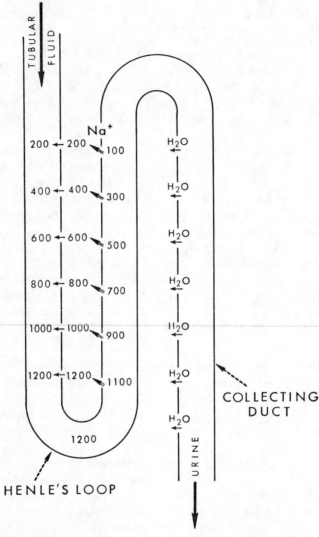

Figure 42. Diagram of the countercurrent multiplier of Henle's loop in the mammalian kidney. The figures refer to sodium concentrations, arbitrarily chosen to show the increase in concentration towards the bottom of the loop. Double slanted arrows indicate active ('up-hill') transport, single arrows passive diffusion. Sodium is actively transported out of the ascending leg of the loop, diffuses into the descending leg, and thus accumulates in the bottom of the loop and in the interstitial tissue. Water is then withdrawn osmotically from the collecting ducts, leaving a concentrated urine.

and can serve to withdraw water osmotically from the fluid in the collecting ducts (figure 42).

For reasons which we need not discuss in detail, the maximum concentration that can be achieved in the renal multiplier depends on the length of the hairpin loop; what we need to emphasize is that we have a loop in which fluid in the two legs flows in opposite directions, and that the function of the countercurrent multiplier system completely depends on two prerequisites; (*a*) there must be an active, energy-requiring transport, and (*b*) the fluid must flow in a hairpin loop for the multiplying effect to occur.

The swimbladder. A fish that lives at a depth of, say, 1000 m is exposed to the full pressure of the surrounding water, in this case 100 atmospheres. The gas in the soft-walled swimbladder is at this same pressure, just as the gas in an air-filled rubber balloon would be if we took it down to the same depth. The surrounding water at this depth, however, has gases dissolved in it only to a total tension of 1 atmosphere, for any given volume of water that has come into equilibrium with atmospheric air at the surface will still contain the same dissolved gases if moved down to our depth of 1000 meters.

The fish then has two problems. First, to fill its swimbladder it must move gases from a tension of about 1 atmosphere in the surrounding water to the much higher pressure inside its swimbladder (in our case a 100-fold increase), obviously an energy-requiring active transport. Secondly, once the gas is in the swimbladder at high pressure, it must be kept from diffusing into the blood, which in the gill comes into diffusion equilibrium with the low gas tensions of the surrounding water.

As always, to understand function we must have some knowledge of structure. In the wall of the swimbladder of most fishes there is a gas gland, which is supplied with blood through a peculiar structure known as the *rete mirabile* (plural, *retia mirabilia*). The form of the rete and the gas gland varies a great deal from species to species, but in principle it is as follows: the artery supplying the gas gland splits up into an enormous number of straight, parallel capillaries, these unite again into a vessel that supplies the gas gland. The vein returning from the gas gland likewise splits up into a set of parallel capillaries, interspersed between the arterial capillaries with amazing regularity, before they again unite and form a single, normal vein.

The capillaries in the rete are quite long, in some deep-sea fish as long as 25 mm. This is an enormous length for a capillary; muscle capillaries which otherwise are amongst the longest anywhere are only of the order of 0.5 mm long. (Those fish which secrete gas under the highest pressures seem to have the longest capillaries, and this makes sense, for the effectiveness of the multiplier is related to its length.) In diagram form the swimbladder with its rete looks somewhat like figure 43, and it is well worth-while to

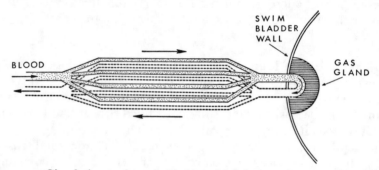

Figure 43. Circulation to the swimbladder of fish is through a rete mirabile in which long arterial capillaries are interspersed between similar veins.

examine the quantitative aspects of the structure. This was done by Krogh (1929), who examined the two parallel retia of a 'medium-sized' eel, which in Denmark would mean a few hundred grams. The retia consisted of 4 mm long capillaries, and had a combined volume of 64 mm³ (or μl), about the size of a drop of water. There were over 100 000 arterial capillaries and nearly as many venous ones, and this gives a total length of 400 meters of each kind of capillary. Their diameter was, as most capillaries, some 7 to 10 μm, and simple arithmetic gives as their total wall surface the impressive area of 100 cm² for each kind, all within the volume of a water drop! For our purpose the important fact is that this enormous area is available for diffusion.

Let us now first look at the problem of keeping gas inside the swimbladder once it has been placed there by a secretory process that we shall deal with later. The situation is completely analogous to the flipper of the whale. Blood coming from the swimbladder will have gases dissolved at high pressure, but as it flows in the rete it meets arterial blood that is lower in gas content. Due to the large surface area available for diffusion, all the excess gas can diffuse across to the incoming arterial blood, and the venous blood will exit from the rete with gas tensions reduced to that of the incoming arterial blood. In spite of the high pressure of the gas in the swimbladder and the low tension in the arterial blood, the gas remains in the swimbladder; the rete serves as a lock or gate past which the gas cannot escape.

When it comes to putting gas into the swimbladder, it is simplest to deal with oxygen first, for this gas is most commonly secreted into the swimbladder. In figure 44 the rete structure has been

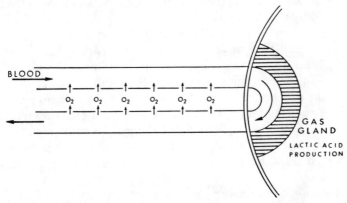

Figure 44. Diagram of the circulation to the swimbladder. Lactic acid produced by the gas gland increases the oxygen tension in the blood which leaves the gland. Oxygen therefore diffuses from the venous capillary to the arterial capillary, and oxygen gradually accumulates in the loop. The reaction rate for the hemoglobin–oxygen binding is an important factor, as explained in the text.

reduced to a single loop so that we have a simple diagram at hand. Let us now assume that the secretory process begins with the gas gland producing lactic acid (which in actuality it does). This lactic acid enters the blood and reduces its affinity for oxygen. The effect of acid on hemoglobin is known as the Bohr effect, but in most fish blood the effect is much more pronounced than in mammals and is known as the Root effect (Scholander and Van Dam, 1954). The result of the acid is to drive oxygen off the hemoglobin, and the concentration of oxygen dissolved in the blood (i.e. not bound to hemoglobin) is therefore increased in the venous blood leaving the gland. Oxygen therefore diffuses across to the arterial capillary where the concentration is lower, and thus remains in the loop where it accumulates. More and more oxygen will accumulate as long as the venous blood leaves with less oxygen than that contained in the incoming arterial blood. Owing to the countercurrent flow lactic acid will also tend to remain in the loop, increasing the effect. Of course, for oxygen to accumulate in the loop the total *amount* of oxygen in the venous blood as it leaves the rete must be less than in the incoming arterial blood; on the other hand, for oxygen to diffuse from the venous to the arterial capillary its *tension* must be higher in the venous capillary.

One factor that contributes importantly to this state of affairs is related to the reaction kinetics of the hemoglobin–oxygen binding as clarified by the Norwegian investigator, Steen. As acid is added to blood, oxygen is driven off the hemoglobin; the reaction is very fast, with a half-time of 50 msec. In contrast, the reaction in the other direction, binding of oxygen to hemoglobin, is a slow process with a half-time of some 10 to 20 sec (Berg and Steen, 1967). This means that the oxygen that was driven off the hemoglobin in the loop does not have time to recombine with the hemoglobin during the short time the blood flows along the venous capillary and out; a high oxygen tension therefore persists in the venous capillary, and oxygen can diffuse across to the arterial capillary and remain in the loop.

By constantly returning oxygen to the arterial capillary, the loop gradually accumulates oxygen, which can reach very high concentrations. How the gas gland itself functions is not known in detail, but at least we now know how very high oxygen concentrations can be built up in the blood that enters the gland.

In addition to acid, the tension of oxygen can be increased by other solutes; the solubility of oxygen (and other gases in solution) is decreased due to a 'salting-out' effect. Since nitrogen can also be secreted into the swimbladder, and in some fishes seems to be the primary gas secreted, the salting-out effect could work in the rete in a way similar to that described above for oxygen (Kuhn *et al.* 1963). This is very helpful, for although we cannot invoke the acid effect on hemoglobin as a means for concentrating inert gases, we still have a means of using the rete as a countercurrent multiplier to build a small effect into a high gas pressure.

To summarize, the countercurrent multiplier responsible for secreting gas under high pressure into the swimbladder has the two prerequisites for a multiplier: (*a*) overall, the secretion is against a concentration gradient or 'up-hill', and (*b*) it depends on a loop structure for the multiplication to take effect.

Whale arteries and an unlikely idea

Not all vascular networks or retia have to be countercurrent exchangers. First of all, two streams are required before we can even begin to look for an exchange. Many whales and dolphins have extensive, dense vascular networks connected with the arteries leading to the skull and along the anterior part of the vertebral column. It is meaningless to look for a countercurrent exchanger in this system, for the arteries are not interspersed with a similar system of veins.

Anatomically, the networks are quite well described, but their function has not been much studied. Nevertheless, some general statements can be made. Diving mammals have a very large blood volume and a very high oxygen capacity of the blood as well, thus giving a large pool of oxygen available during the dive. During a dive, the blood flow to much of the body is reduced to zero and the muscles in particular are virtually bloodless. The large volume of blood must therefore be stored elsewhere in suitable volume receptacles.

Such receptacles preferably should contain arterialized blood, not venous, for the larger the pool of oxygen, the longer the possible dive. The best place to store blood is in small vessels with elastic walls, rather than in big, flaccid sinuses, for small vessels will better help to maintain the arterial pressures during the long period between heart beats. Thus, the most vital tissues, the central nervous system and the heart muscle, are supplied with oxygen-containing blood under maintained arterial pressure, although the heart rate may be down to a few beats per minute.

A rather ingenious (but in fact unlikely) interpretation of the arterial networks of the whales is that they help during a dive in utilizing oxygen stored in the surrounding fat, for the solubility of oxygen in fats and oils is some five times higher than that in water. A look at the quantities of oxygen involved is helpful. The solubility of oxygen in saline, at the arterial oxygen tension (about 100 mm Hg), is about 0.3 ml O_2 per 100 ml, in fat it is about 1.5 ml O_2 per 100 ml. This is not overwhelming in comparison to the oxygen capacity of blood, which is 20–30 ml per 100 ml, or some 20-fold that in oil. A little blood can hold more oxygen than a lot of fat! And remember that the size of the vessels we are talking about is in the millimeter range, 100-fold larger than that needed for reasonably fast gas diffusion.

But what about adaptation to pressure? Whales have interesting anatomical features, some are obvious adaptations to swimming and diving, the function of others is more obscure. It is tempting to interpret any and all unusual features found in a whale as adaptations, if not to swimming and diving *per se*, then to the high pressures encountered during dives to great depths. Admittedly, whales have structures that we do not understand, but before we explain them we must somehow be able to suggest how they work. An hypothesis is not merely to say that we believe, there ought to be some reasoning behind it.

Let us again consider the large arterial networks in whales, which have been said to aid in resisting the pressures encountered during deep dives. It has been suggested that they protect the brain, but how can this be achieved? We know that in a fluid, pressure is distributed evenly in all directions, and only a strong rigid structure can resist great pressures. When a whale dives, its soft tissues, blood-vessels, blood, etc., are at virtually the same pressure as that of the surrounding water.* The brain is enclosed in a rigid case, the skull, which in whales is very strong and dense. To keep a lower pressure in the brain would therefore be feasible, provided that all openings into the skull were closed; however, if the blood vessels are open, the pressures will immediately equalize. To maintain a lower pressure inside the skull will therefore require, not only that circulation to the brain is stopped, but also that valves of some sort keep the blood from flowing in, valves strong enough to support the postulated pressure differences. If the brain does receive blood during a deep dive, and everybody now agrees that this is so, the pressure inside the skull must equal that outside.

This is not to say that the arterial retial structures along the vertebral column and at the base of skull of whales cannot somehow be related to diving and pressure; but a postulated hypothesis must include not only a reasonable suggestion of what function it should serve, but also *how* the proposed function is physically possible.

There is another unlikely idea I should like to mention. It has been suggested that the rete structures in the upper arm of some climbing animals, e.g. the sloths and lorises, and especially pro-

---

* We can expect that the usual arterial pressure needed to circulate the blood is maintained, but this is a minute magnitude compared to the hydrostatic pressure of a deep dive.

nounced in the potto, are responsible for the powerful prehensile grasp these animals can maintain for long periods.

The authors of such claims by no means make it clear how a vascular bundle in the upper arm should contribute to the ability of the underarm muscles to maintain a strong and prolonged contraction. Below the rete the vessels are again united into a single artery; how should the muscles know that the blood they receive from this artery should somehow differ from that coming from any other artery? The only difference could be a slightly lower pressure.

One suggestion is that the veins in the rete, because they are interspersed between the arteries, will remain open when external pressure increases during the strong muscular contraction, while ordinary veins would collapse (Suckling, Suckling and Walker, 1969). But what about the veins just below the vascular bundle, if they collapse the rete would be of little use. Anyway, the rete is in the upper, and the contracting muscles in the lower arm. It would be helpful if hypotheses of this kind could give a clear explanation of how the proposed mechanism would work, and preferably also be supported by experiments showing that it *does* work. What remains is that heat exchange in the rete is inescapable, and also that venous blood returning from a limb dipped in ice-water is warmed to near-arterial temperature after it has passed through the rete (Scholander and Krog, 1957). That these retia are heat exchangers is beyond doubt; what other function they may have, if any, needs better substantiation.

In search of generalizations

I have already touched on most of the general features of the various countercurrent mechanisms which I have mentioned. Nevertheless, at the risk of being intolerably redundant, I shall again set out the main points.

(1) *Countercurrent multipliers* work 'up-hill', they require energy, and they depend on a loop structure in which a solute is accumulated. The channels of flow are of capillary size; known examples include formation of a concentrated urine in the kidney and gas accumulation in the swimbladder.

(2) *Countercurrent exchangers* work 'down-hill', they work as locking-in devices to preserve gradients of heat, gases, etc. A loop structure is not needed, but the flows must be in opposite directions.

There may be different fluids in the two channels (water–blood in the fish gill; air–blood in the bird lung; blood–blood in the placenta), but there *may* be a loop with the same fluid (blood to the whale flipper, the testes, the brain).

Exchangers work only for diffusible substances (heat, oxygen, water vapor). Down-hill flow of these substances cannot be prevented; the efficiency will depend on area, diffusion distances, and other parameters (rates of flow, heat capacity or gas solubility, etc.).

Heat exchangers are efficient when the size of the blood vessels is in the range of 1 mm, in this size vessel gas exchange is minimal. Gas exchangers depend on blood vessels of capillary size, about 0.01 mm, or on diffusion in air which is 100 000 times faster than in water.

6. Body size and problems of scaling*

Comparative physiology is based on the premise that animals are more or less similar and thus can be compared. This does not mean that they are alike, and a deviation from the general pattern is often as meaningful and interesting as the similarities.

The difficulty of comparing organisms of different size is symbolized in figure 45, which shows our old friend Gulliver, as

Figure 45. Gulliver presented a problem of scaling, the Lilliputian emperor had to decide how much food the Man Mountain needed. (Reproduced by permission of Doubleday & Company, Inc. from *Gulliver's Travels* by Jonathan Swift, illustrated by Leonard Weisgard. Illustrations copyright 1954 by Nelson Doubleday, Inc.)

---

* This chapter was prepared as the introductory lecture to a symposium, *Energy Metabolism and its Regulation*, held in honor of Max Kleiber at Davis, California on 29 August 1969. It was published in *Federation Proceedings* 29, 1524-32, 1970, and is reprinted, with minor changes, with the permission of the publishers.

well as a Lilliputian walking down the cobblestone street. The immediate problem that the Lilliputian emperor had to face when Gulliver arrived was how much food the Man Mountain needed. Swift (1726) reported that it was exactly 1728 Lilliputian portions. Is this estimate correct, and on what basis can we compare mammals of different size?

Those who have dissected a racehorse or a greyhound may have noted that these animals have larger than proportionate hearts. In proportion to their body size mammals generally have very similar heart size, about 5 or 6 gram per kg body weight, and we are so used to this scale that we immediately notice a deviation.

The fact that most mammals are similar in that they have just above 0.5% of their body weight as heart, may, at first glance, be surprising, for we know that the small animal in relation to its body size has a far higher metabolic rate than the large one. To supply the tissues with oxygen at the necessary high rate obviously cannot be achieved by merely adjusting stroke volume, which is limited by the size of the heart; thus, the heart frequency remains the major variable to adjust, as evident from heart rates between 500 and 1000 per minute in the smallest mammals.

LSD and the elephant

Among mammals in general, perhaps the most conspicuous difference is their size. A 4 ton elephant is a million times as large as a 4 gram shrew, and the largest living mammal, the blue whale, can be another 25-fold larger. Twenty-five million shrews put together are very difficult to imagine, and since they eat about half their body weight of food per day, they would be very hard to feed. We can conceive of twenty-five elephants put together, although it is a formidable mass of elephants.

If we just consider the size of an elephant, we can easily make some serious mistakes. A few years ago, the journal *Science* published a note which described the reaction of a male elephant to LSD. The investigators wanted to study the peculiar condition of the male elephant known as 'musth'. A male elephant in 'musth', a word of Sanscrit origin, is violent and uncontrollable, but he is not in rut (and, although some people think so, musth doesn't mean that he 'must' have a mate). Shortly after the publication of this note, a letter to the editor of *Science* described the calculation of the dose of LSD as 'an elephantine fallacy' (Harwood, 1963). The authors had calculated the dose, based on the amount that puts a cat into a rage, and had multiplied up by weight until they arrived at 297 mg of LSD to be given to the elephant. The long description of what happened can be shortened by saying that after the injection of the 297 mg (enough for 1500 'trips', for a single human dose is about 0.2 mg), the elephant immediately started trumpeting and running around, then he stopped and swayed, five minutes after the injection he collapsed, went into convulsions, defecated, and died. The authors concluded that elephants are peculiarly sensitive to LSD.

I am using this example to illustrate the tragic events that may result from a lack of appreciation of the problems of scaling. How should we calculate drug dosage? Of course, if we want to achieve equal concentrations in the body fluids of a small and a large animal, we should calculate in simple proportion to their weights. If this calculation is done by extrapolation from a 2.6 kg cat and its dose of 0.1 mg LSD per kg, we arrive at the now obviously lethal dose of 297 mg of LSD for the elephant. If instead, as the letter-writer in *Science* said, we calculate on the basis of metabolic rate, we find that a much smaller dose of 80 mg is needed. This makes some sense, for we can expect that detoxification of a drug or its excretion may be related to metabolic rate. But there could be other considerations or special circumstances. For example, LSD could be concentrated in the brain, and in that event we would have a much more complex situation and might want to consider brain weight.

We could also use as a basis for the calculation an animal which is not as notoriously tolerant to LSD as cats, for example, we could use man. The weight of a man is 70 kg, and a dose of only 0.2 mg LSD gives him severe psychotic symptoms. On a weight basis this suggests that the elephant should receive 8 instead of nearly 300 mg LSD. Based on metabolic rate, which in man is $13 \ 1 O_2 \ h^{-1}$ and in the elephant, 210, we get a 3 mg dose for the elephant. If we consider the brain size, which in man is 1400 g and in the elephant about 3000 g, we arrive at only 0.4 mg (see table 6).

I don't intend to say how much LSD should have been injected, if any; I only wish to point out that scaling obviously is not a simple problem. There are, however, many situations in which deviations from strict similarity are more obvious and easier to analyze.

Table 6. *Possible calculations of a 'suitable' dose of LSD to give to an elephant*

(a)	Based on body weight and dose effective in cats	297 mg
(b)	Based on metabolic rates of elephant and cat	80 mg
(c)	Based on body weight of elephant and dose effective in man	8 mg
(d)	Based on metabolic rates of elephant and man	3 mg
(e)	Based on brain size of elephant and man	0.4 mg

Scaling the skeleton

Figure 46 shows two animals that none of us has ever seen alive. To the right is a *Neohipparion*, an extinct ancestor of the horse, and we can immediately see that this must be an animal the size of a small deer. To the left is another extinct animal, the *Mastodon*,

Figure 46. Two extinct mammals, *Neohipparion* (*right*) and *Mastodon* (*left*). Skeletons must be scaled to support the weight of the body as its bulk increases with the third power of linear dimensions. (From Gregory, 1912.)

drawn to the same size, and yet we at once know that these large and heavy bones must be those of an animal the size of an elephant. Our instantaneous perception of the true size of these animals, although drawn to equal size, expresses a well-known rule of scaling. The mass of an animal increases as the third power of the linear dimension, and, to support this mass, the cross-section of the bones must be increased beyond what is achieved by increasing their diameter in linear proportion. Linear scaling would just square the supporting cross-section of the bone, and the bones would be crushed by the weight of the cubed increase in mass.

The increase in skeletal weight with the body size of mammals is expressed in a more general way in figure 47. This graph is plotted on logarithmic coordinates, and the points fall on a straight line. If the increase in all body dimensions were scaled linearly, the weight of the skeleton would remain as the same percentage of the body weight, and we would have a line with a slope of 1.0. Instead, we find a line with a steeper slope, the large animals have relatively heavier skeletons and the slope is 1.13.

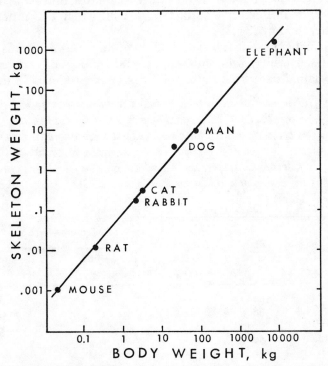

Figure 47. Weight of the skeleton of mammals increases more than proportionally to an increase in body weight. The slope of the regression line is 1.13. (Data from Kayser and Heusner, 1964.)

Metabolic rate, size, and surface

A physiological variable which concerns us much more than the skeletal size of mammals is their metabolic rate. If metabolic rate were scaled directly relative to the body weight we would, of course, find a regression line with a slope of 1.0 in a log–log plot. The fact that this is not possible is so well known that I hesitate to restate it, but it will be necessary for my further discussion. More than 100 years ago French scientists pointed out that heat dissipation from warm-blooded animals must be proportional to their free surface, and small animals must, because of their larger relative surface, have a higher relative rate of heat production than large animals. Similar considerations led to the formulation of Bergmann's rule, a rule which claims that animals in colder climates are usually larger than close relatives from warmer climates, thus reducing the relative area of external surface from which they can suffer heat loss (Bergmann, 1847).

The first experimental examination of these problems was made by Rubner (1883), who published a study of dogs of various sizes. The original data from Rubner's work are reproduced in table 7.

Table 7. *The earliest study which related metabolic rate to body surface rather than to weight, was based on a study of dogs. Although there is better correlation with surface than with weight, it is now clear that body surface per se does not determine metabolic rate. Nevertheless, body surface constitutes a constraint on what is possible in the design of a well-functioning animal.* (Data from Rubner, 1883)

| Body weight kg | Body surface cm² | Surface: Weight cm² kg⁻¹ | Metabolic rate | |
			per weight kcal (kg day)⁻¹	per surface kcal (m² day)⁻¹
31.20	10750	344	35.68	1036
24.00	8805	366	40.91	1112
19.80	7500	379	45.87	1207
18.20	7662	421	46.20	1097
9.61	5286	550	65.16	1183
6.50	3724	573	66.07	1153
3.19	2423	726	88.07	1212

The first column shows that the dogs ranged from 31 to about 3 kg, a size range of 10:1. (Today, we could easily find dogs of much more different sizes.) In these dogs the metabolic rate per kg (column 4) increased as the body weight diminished. On the other hand, if the metabolic rate was calculated per body surface area (last column) there was a nearly constant relation between metabolic rate and body surface of the dogs. Rubner interpreted this as necessary for the animal to keep warm, for heat loss takes place from the surface and must be related to the extent of this surface. He also very explicitly stated that this is not due to any specific activity of the cells, but that it is due to the stimulation of receptors in the skin which in turn act on the cells of the metabolizing tissues. We now consider this argument erroneous.

These findings established the 'surface law', or the 'surface rule' as I prefer to call it, and the extensive use of body surface as a base of reference for metabolic rate. It became evident, however, only five years after Rubner's work, that the need for heat dissipation cannot be the primary reason for the relationship. In 1888, von Hoesslin published a study of fish, and he found that in these also, oxygen consumption showed a closer relation to surface than to weight. Obviously, fish have little need to keep warm in proportion to the extent of their body surface.

Since that time there have been innumerable studies of this subject, but before I discuss these, I should like to give some further attention to the body surface of animals. The body surface area is a surprisingly regular function of body size. On the other hand, it is difficult to determine with accuracy the exact size of the surface area, and Kleiber (1967) has repeatedly pointed out that an accuracy of better than 20% cannot by any means be expected.

Surface areas of a large number of organisms, ranging in weight from a few grams to several tons, are compiled in figure 48. In this graph the fully-drawn straight line indicates the surface area of a sphere of a given weight and a specific gravity of 1.0. The sphere, of course, has the smallest possible surface of any geometrical shape of a given volume, and compared to this we see that animals have surface areas that quite regularly amount to about twice the areas of spheres of the same weight. Someone may ask about the deviating points in the upper right hand of the graph – they are not animals, they are merely beech trees, and the reason for including these should become evident later on. The trees have larger relative surfaces than animals, but they are on a line parallel to that of animals, which has a slope of 0.67. In general, then, the surface of living organisms seems to be an amazingly regular function of the square of their linear dimension (or the two-thirds power of their mass or volume).

It is now important to realize that not only for geometrical reasons is it difficult to deviate from the regularity of the relationship of surface to volume, but if we wish to design a workable animal, large or small, we find ourselves inextricably enmeshed in the need for considering surfaces. A moment's thought gives us a long list of surface-related processes, heat loss, as we discussed before, must be surface related, the uptake of oxygen in lungs or in gills depends on the area of their surface, the diffusion of oxygen through the walls of capillaries must be related to their surface areas, food uptake in the intestine likewise must be surface related, and so on. In fact, all cells have surfaces, and surface processes, or membrane processes as we call them now, must be related to the areas of these surfaces.

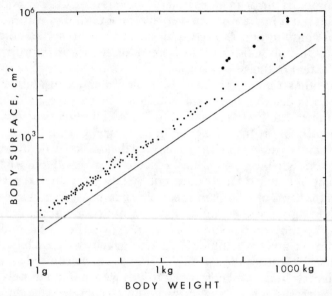

Figure 48. Body surface of vertebrates in relation to body weight. The fully drawn line gives the surface of a sphere with a density of 1.0. The larger points in the upper right-hand corner represent beech trees (Data from Hemmingsen, 1960.)

Surface, a stifling constraint, or
Easy equations

Obviously, we could hardly design an animal with disregard to surface relationships. To mention an example from problems of temperature regulation, Kleiber has stated that if a steer is designed with the metabolic rate of a mouse, to dissipate heat at the rate it is produced, its surface temperature would have to be well above the boiling point. Conversely, if a mouse is designed with the weight-related metabolic rate of a steer, to keep warm it would need to have as surface insulation a fur at least 20 cm thick.

Obviously surface considerations are essential and cannot be ignored.

After Rubner's work, several eminent physiologists gave a great deal of attention to the relation between body surface and metabolic rate. The basic thinking in the field, however, remained confined by undue emphasis on surface areas, until Kleiber (1932) wrote an article that has ever since influenced all our concepts of the subject. He published his paper in *Hilgardia*, a little-known journal of agricultural science published by the California Agricultural Experiment Station at Davis, now the University of California at Davis. I have a rare reprint of this paper, it bears a red stamp which says 'Max Kleiber, Personal Copy, Do Not Remove From Files'. Kleiber has obviously realized the logical fallacy of these directions, for if it is not removed from the files, how can it be used? This must be the reason that he personally gave it to me, a gift for which I am very grateful indeed.

In this paper Kleiber showed that the metabolic rate of mammals not only is an amazingly regular function of their body size, but also that it is significantly different from a direct surface function. He examined animals over a size range from rats to steers, and published a curve that could be called a 'rat-to-steer curve'. The phrase doesn't form easily in the mouth and hasn't become well known; what is recognized, however, is Kleiber's generalization that on a log–log scale the metabolism of mammals as related to their body weight forms a straight line with a slope of 0.75. Two years later Brody, Procter and Ashworth (1934) published their well-known 'mouse-to-elephant curve', and four years later again Benedict (1938) published a similar curve in his book *Vital Energetics* in which he somewhat reluctantly admitted that, although animals do not know about logarithms, in a log–log plot the points fall amazingly close to a straight line.

The slope that Brody found was 0.734, very close to 0.75 which Kleiber had suggested should be used, in fact, the difference between these numbers is not statistically significant. Quibbling about the validity of the second and third decimal of the exponent appears somewhat meaningless when we realize that before Brody calculated his line he made certain 'adjustments' for the elephant. A point at the extreme of the size range influences the slope more than would a point in the middle, and in this case it consisted in '30% deducted from the original value (10% for standing & 20% for heat increment of feeding)' (Brody, 1945, p. 388).

Similar studies, relating metabolic rate to body size, have been extended to numerous cold-blooded vertebrates, and, although the regression lines are lower, their slopes are similar. Even the metabolic rates of beech trees fall on a line with a similar slope. This should forever dispel the notion that temperature regulation is a primary stimulus in the regular relationship between heat production and body size. Likewise, many invertebrates have metabolic rate–body-size curves on similar straight lines, which within statistical limits can easily be said to have the same slope. Some, however, have significantly different slopes, and a few, some snails and some insects, have been found to have metabolic lines with slopes of 1.0, that is, their metabolic rate is directly proportional to the body weight.

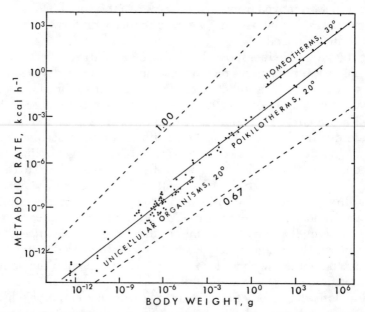

Figure 49. Metabolic rates of various organisms in relation to body weight. Note that each mark on the coordinates denotes a 1000-fold difference in magnitude. (Data from Hemmingsen, 1960.)

The entire field has been reviewed by Hemmingsen (1960) in a monumental summary which covers the range from the smallest micro-organisms to the largest known mammals (figure 49). The wide range covered is reflected in the fact that each division on the coordinates of figure 49 stands, not for a 10-fold, but a 1000-fold difference. Warm-blooded animals fall on a very nice straight line, and cold-blooded vertebrates on a similar line but at a lower level which continues down through many invertebrates. Micro-organisms again seem to be organized on a line with a similar slope. The slope is significantly different from a surface relation (indicated by the line marked 0.67) or a direct weight relationship (represented by the line marked 1.00).

The equation which describes these lines is the familiar exponential equation:

$$y = b \cdot x^a$$

In the logarithmic form this equation gives a linear function:

$$\log y = a \cdot \log x + \log b$$

where a, the slope of the straight line, is the exponent in the preceding equation. Similarly, when we refer to the metabolic curves we have been discussing, we have log metabolic rate related to log body weight by a straight line with the slope a, or:

$$\log \text{ metabolic rate} = a \cdot \log \text{ body weight} + k$$

I should now like to return to the biological meaning of this exponent a, or the slopes of the straight regression lines, and the statistical significance of differences between them. Is the slope 0.75 as proposed by Kleiber significantly different from 0.67? The answer is that to establish a significant difference between these requires animals which differ in size by more than 9 to 1. We now remember that this is approximately the size range of Rubner's dogs, and therefore Rubner really did not have material to establish with certainty a slope different from 0.67, or a simple surface relationship. In those studies that include a large number of mammalian species of widely different sizes, exponents have consistently been higher than 0.67. For example, Brody and Procter (1932) derived 0.734 and Kleiber 0.756. Kleiber (1961) has shown that these exponents cannot be established as significantly different from 0.75 unless we examine animals that cover a range from smaller than 4 g up to larger than 800 tons. The blue whale, the largest living animal, may weigh 100 tons, and I therefore imagine that we will never have experimental observations to support a slope significantly different from 0.75. One reason that Kleiber advocates the use of the simple $\frac{3}{4}$ power is the much greater simplicity in the arithmetical computations that this permits.

We can now summarize the events that led up to the present-day concepts of this field. Early in the last century French writers realized that heat loss must be related to the free surface of the animal, whatever that is. Rubner made the first experimental study of the problem, using dogs of widely different sizes. He did indeed find a close relation to body surface, and thus established the body surface as a common reference point in metabolic studies. Kleiber after examining metabolic data from mammals over a much wider size range, stated that the exponent is much closer to the $\frac{3}{4}$ power of the body weight.

In the 1967 volume of *Annual Review of Physiology*, Kleiber humorously emphasized the universality of this exponent. In a prefatory chapter, entitled 'An old professor of animal husbandry ruminates', he discussed several interesting subjects, including Lilliputian physiology. After making some assumptions about their size, he found that the Lilliputians had used the slope 0.76 when they calculated Gulliver's need for food; that is, a slope not significantly different from that in the *Hilgardia* paper in 1932, and that the Lilliputians thus had anticipated him by 233 years. Kleiber also pointed out that if they had used Rubner's surface rule of 1883, Gulliver would have received only 675 portions and would have starved miserably.*

Oxygen supply, more scaling

Can we analyze what the very common slope of $\frac{3}{4}$ really means? I should like to look at some of the difficulties encountered in the scaling of an organism before I try to answer this question. The way metabolic rate is plotted in figure 50 gives us a better visual impression of how rapidly metabolic rate increases with decreasing body size. The metabolic rate, when calculated per gram body weight and plotted on a linear scale, points out the tremendous increase in the small animal. This again means that we must supply to the cells of the smallest mammal, oxygen and nutrients at rates that are some 100-fold as great as in the largest mammal. I shall therefore discuss for a moment the parameters that govern the oxygen

---

* A careful reading of *Gulliver's Travels* will reveal some discrepancy between Kleiber's assumptions and Swift's statements in regard to the size of Lilliputians. Actually, Swift was not well versed in modern physiology and scaled the ration according to weight, calculated as the third power of the linear dimension.

supply to the tissues, and afterwards consider the gas exchange in the lungs, both being essential components in the oxygen supply.

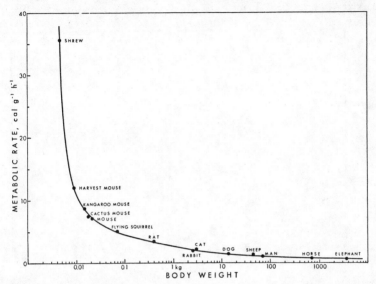

Figure 50. Observed metabolic rates of mammals, calculated per gram body weight and plotted on an arithmetic ordinate. This plot shows the tremendous increase in the metabolic rate of the very small mammals.

Capillary density. The rate of diffusion of oxygen from the capillary to the metabolizing cell is determined by (*a*) the diffusion distance, and (*b*) the diffusion head (or difference in P_{O_2} from capillary to cell). The former is simply a function of capillary density, while the latter is influenced by the dissociation curve for blood and implemented by the Bohr effect, the Bohr effect in turn needing the presence of carbonic anhydrase to be effective within the short time the blood remains in the capillary.

Krogh (1929) was aware of the need for a higher capillary density in small animals, and he confirmed this by quantitative determinations of capillary numbers in muscle from horse, dog, and guinea pig. The same general trend has been reported by several later investigators, but differences in technique make comparisons between species uncertain. We therefore decided to use uniform techniques and examine several different muscles from a large number of mammals. We found wide variations from muscle to muscle within one animal which were only in part related to the proportion of red and white muscle fibers, and a much less consistent body size relationship than had been assumed to exist (figure 51). True enough, the smallest animals had very

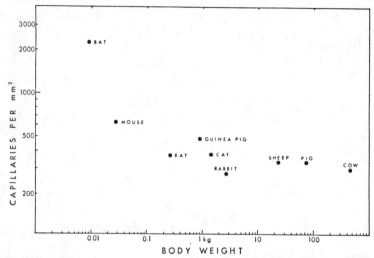

Figure 51. Capillary density in the gastrocnemius muscle. In very small mammals the capillary counts are high, but for a body size of a rat or larger there seems to be no clear trend in relation to body size. (Data from Schmidt-Nielsen and Pennycuik, 1961.)

high capillary densities, but the larger ones, from the rabbit up, displayed no certain trend in their capillary densities. As far as I can see, this should mean that other scaling considerations are more important than diffusion distance, probably for entirely different reasons. The muscle capillaries are, of course, interspersed by muscle fibers, and in any event the muscle fiber is the primary functional element of the muscle. Perhaps it is impossible to scale up the size of a muscle fiber in a large animal, maybe for reasons related to the size of the contractile elements or to the conduction system for the action potential. If this is so, there will be constraints on how far capillary density can be changed in the large animal's muscles.

Unloading oxygen. Let us next look at the unloading tension for oxygen in the capillary blood, as expressed by the dissociation curve (figure 52). It is necessary that for these purposes we compare whole, unaltered blood at the normal pH of the organism, for it is whole blood, and not dilute hemoglobin solutions in phosphate buffer, that runs in our blood vessels. The trend is unmistakable, the large animals have dissociation curves to the left and small animals to the right. We often refer to the half-saturation pressure (P_{50}) as the 'unloading tension', but we do not have to establish

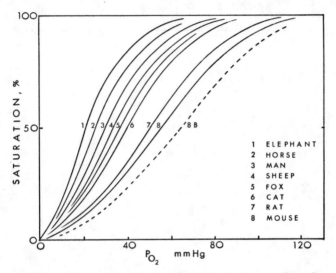

Figure 52. Dissociation curves of oxyhemoglobin from a number of mammals show that the blood of smaller mammals will unload oxygen under a higher oxygen pressure than that of larger mammals. This helps in the delivery of sufficient oxygen to the tissues to maintain the high metabolic rates of small animals. The dashed curve (8B) indicates the effect of acid (Bohr effect) on mouse blood (curve 8). (Data from Schmidt-Nielsen and Larimer, 1958. Elephant from Bartels *et al.* 1963.)

the exact unloading tension to see that in the smaller animal the hemoglobin gives up its oxygen at a higher oxygen pressure than in the larger one. Figure 52 also shows the effect of acid on the dissociation curve for the mouse, indicated by the dotted curve to the right of the normal curve for the same animal. This shift to the right when acid is added (Bohr effect) increases the unloading tension for oxygen.

After the relationship between dissociation curve and body size of mammals had been pointed out and interpreted as being related to the greater need for oxygen in the small mammal, there has been an increased interest in studies of whole blood, and additional mammals have been found to adhere to the same general pattern.

Of particular interest because of its size is the elephant, which has a curve to the left of all the other mammals.

The Bohr effect. As mentioned a moment ago, the Bohr effect is of great help because it increases the unloading pressure for oxygen. Figure 53 shows a compilation made by Riggs (1960) of

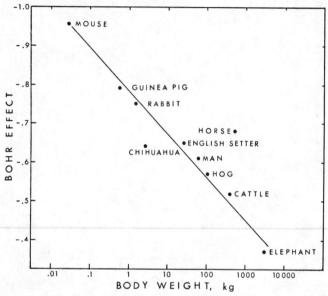

Figure 53. The effect of acid on the unloading of oxygen from the blood (Bohr effect, expressed as $\Delta \log P_{50}/\Delta$pH) is greater in small than in large animals. (Data from Riggs, 1960.)

the Bohr effects of various mammals. He found that from the elephant to the mouse there is a great increase in the magnitude of the Bohr effect; acidification of the blood contributes more to the unloading of oxygen in the smaller animal. The chihuahua is interesting, it deviates considerably from the regression line and thus seems anomalous. Perhaps the fact that its Bohr effect is the same as that of a normal dog indicates that in spite of its half-kilogram size and peculiar appearance, the chihuahua still is a dog.

Carbonic anhydrase. In order to utilize the Bohr effect for an increase in oxygen delivery, it is quite necessary for the blood to be acidified before it leaves the capillary. The time a red cell remains in the capillary is only a fraction of a second, and without carbonic anhydrase the CO_2 from the tissues would not be hydrated and form carbonic acid in this short time. Larimer and I have studied the concentration of carbonic anhydrase in the red cells of various mammals, and indeed, the small animals have a significantly higher concentration of this enzyme in their red cells than the large animals. Carbonic anhydrase is not considered to be essential in respiration, and inhibition of its function with Diamox has little effect. Davenport has described it as an enzyme in search of a function. We suggested that, while carbonic anhydrase is not necessary in CO_2 transport, its primary role is to aid in the adequate delivery of oxygen to the tissues, an effect that would be particularly valuable during work and exercise (Larimer and Schmidt-Nielsen, 1960).

Scaling of the lung

I shall now briefly deal with the opposite element in the oxygen transport system, the uptake of oxygen in the lungs. Here the scaling problems are a great deal easier to deal with, and we can therefore carry our analysis further.

First of all, we find that the lung volume in mammals is directly related to the body weight (figure 54). By plotting log lung volume against log body weight we obtain a straight line with a slope that is very close to 1.0; in other words, as a general pattern all mammals are scaled similarly and have a lung volume of about 6.3% of the body weight. As I said before, there are always deviations from such regression lines. These may be due to errors or to variations in methods, but again, some deviations may be particularly interesting and may deserve more attention than the general pattern, once this latter has become clearly known.

If next we look at the diffusion area of the lung, we find that this variable is directly related to the oxygen consumption of animals rather than to their body size (figure 55). This does indeed make physiological sense. Oxygen consumption is, of course, related to body weight by a power of 0.75, and therefore the diffusion area is related likewise. From a consideration of scaling relative to body size, it can thus be concluded that the size of the mammalian lung and its diffusion area are indeed adjusted to the metabolic needs for oxygen.

Another example will show that we can further extend the use of allometric equations and their predictive value. The vital capacity of mammals is very nearly the same function of body size as the lung volume in figure 54, and can be described by the equation

$$\text{Vital capacity} = 0.063 \; B^{1.0}$$

The tidal volume is similarly related to body size,

$$\text{Tidal volume} = 0.0063 \; B^{1.0}$$

If we now divide the tidal volume by the vital capacity, we obtain the dimensionless number 0.1. This predicts that in mammals in

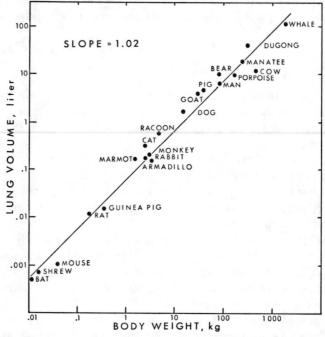

Figure 54. Lung volume of mammals is scaled in simple proportion to body size, as indicated by a slope of the regression line of nearly 1.0. (Data from Tenney and Remmers, 1963.)

general, the tidal volume is one-tenth of the vital capacity, irrespective of their body size. Although there are deviations from this general pattern, the statement has a predictive value which is useful. It has practical use in that we can take a rat or a horse and predict its expected physiological parameters, or we can look at the results of our studies and see how they fit the general pattern, or deviate from it. In this case we ended up with a non-dimensional number which describes the similarity of respiratory ventilation

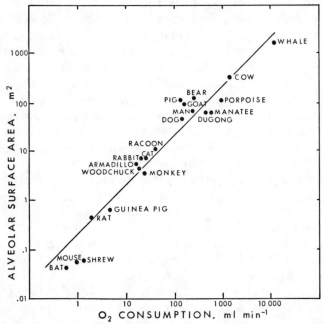

Figure 55. Diffusion area of the mammalian lung is scaled in simple proportion to the rate of oxygen consumption. (Data from Tenney and Remmers, 1963.)

in all mammals. In this regard, therefore, mammals have a simply-scaled similarity without any residual body-size dependent exponent.

Some years ago, Drorbaugh (1960) studied mice, rats, rabbits and dogs in some further detail, and among other parameters he discussed lung compliance. He found compliance to be directly related to the body size (B), expressed as the equation

$$\text{Compliance} = 0.00121 \ B^{1.0} \ \text{ml (cm H}_2\text{O})^{-1}$$

This equation says that the number of ml the lung will be expanded for a change of 1 cm water pressure is directly related to the body weight of the animal, and the lung compliance of a large animal is therefore scaled simply in proportion to its size. The vital capacity, as you remember, is related in the same way to body size. By dividing one equation by the other, we obtain the specific compliance

$$\text{Specific compliance} = 0.019 \ (\text{cm H}_2\text{O})^{-1}$$

In the process we lost the body weight from the equation, and specific compliance therefore is not size dependent. To express this equation in words, for a change in pressure of 1 cm H_2O, all mammals have the same change in lung volume relative to their vital capacity; that is, a change of 1 cm water pressure causes a change of 1.9% in lung volume. The prediction which Drorbaugh correctly made from these considerations is that the pressure per intake of one tidal volume should be the same in all mammals.

The scaling of a large number of physiological variables and the derivation of non-dimensional numbers which describe their inter-relations has become increasingly important. Several years ago Adolph published an important paper in this field (Adolph, 1949), and Walter Stahl, before his death a few years ago, had completed a major monograph on *Physiological Similarity and Modeling* which is now about to be published.

Before I conclude this chapter I find it appropriate to recall that Kleiber initiated the use of exponential equations in analyzing physiological function in relation to body size with his clear treatment of the metabolic rate problem in 1932. We were then relieved of the constraining demand to fit metabolic rate to body surface, and the heated discussions of how to determine 'free' or 'true' surface have therefore subsided. The 'surface law' as such does not even survive as a 'surface rule', but the analysis of function in relation to body size has become an immensely interesting approach in our attempts at finding out how animals work.

References

Adolph, E. F. (1949) Quantitative relations in the physiological constitutions of mammals. *Science, N.Y.* **109**, 579–85.

Albers, C. (1961) Der Mechanismus des Wärmehechelns beim Hund. I. Die Ventilation und die arteriellen Blutgase während des Wärmehechelns. *Pflüger's Arch. ges Physiol.* **274**, 125–47.

Baker, M. A. and Hayward, J. N. (1968) The influence of the nasal mucosa and the carotid rete upon hypothalamic temperature in sheep. *J. Physiol., Lond.* **198**, 561–79.

Banko, W. E. (1960) *The Trumpeter Swan.* North American Fauna, No. 63. 214 pp. Washington, D.C.: U.S. Dept. Interior, Fish and Wildl. Serv.

Barnett, C. H., Harrison, R. J. and Tomlinson, J. D. W. (1958) Variations in the venous system of mammals. *Biol. Rev.* **33**, 442–87.

Bartels, H., Hilpert, P., Barbey, K., Betke, K., Riegel, K., Lang, E. M. and Metcalfe, J. (1963) Respiratory functions of blood of the yak, llama, camel, Dybowski deer, and African elephant. *Am. J. Physiol.* **205**, 331–6.

Benedict, F. G. (1938) *Vital Energetics. A Study in Comparative Basal Metabolism.* 215 pp. Washington, D.C.: Carnegie Inst. of Washington.

Berg, T. and Steen J. B. (1968) The mechanism of oxygen concentration in the swim-bladder of the eel. *J. Physiol., Lond.* **195**, 631–8.

Bergmann, C. (1847) Ueber die Verhältnisse der Wärmeökonomie der Thiere zu ihrer Grösse. *Göttinger Studien,* 595–708.

Brett, J. R. (1965) The relation of size to rate of oxygen consumption and sustained swimming speed of sockeye salmon (*Oncorhynchus nerka*). *J. Fish. Res. Bd. Can.* **22**, 1491–501.

Bretz, W. L. and Schmidt-Nielsen, K. (1970) Patterns of air flow in the duck lung. *Fedn. Proc.* **29**, 662.

Bretz, W. L. and Schmidt-Nielsen, K. (1971) Bird respiration: flow patterns in the duck lung. *J. exp. Biol.* **54**, 103–18.

Brody, S. (1945) *Bioenergetics and Growth. With Special Reference to the Efficiency Complex in Domestic Animals.* New York: Reinhold Pub. Co.

Brody, S. and Procter, R. C. (1932) Relation between basal metabolism and mature body weight in different species of mammals and birds. *Missouri Agr. Exp. Sta. Res. Bull.* **166**, 89–101.

Brody, S., Procter, R. C. and Ashworth, U. S. (1934) Basal metabolism, endogenous nitrogen, creatinine and neutral sulphur excretions as functions of body weight. *Missouri Agr. Exp. Sta. Res. Bull.* **220**, 1–40.

Calder, W. A. and Schmidt-Nielsen, K. (1966) Evaporative cooling and respiratory alkalosis in the pigeon. *Proc. natn. Acad. Sci. U.S.A.* **55**, 750–6

Calder, W. A. and Schmidt-Nielsen, K. (1968) Panting and blood carbon dioxide in birds. *Am. J. Physiol.* **215**, 506–8.

Carey, F. G. and Teal, J. M. (1966) Heat conservation in tuna fish muscle. *Proc. natn. Acad. Sci. U.S.A.* **56**, 1464–9.

Carey, F. G. and Teal, J. M. (1969) Mako and porbeagle: warm-bodied sharks. *Comp. Biochem. Physiol.* **28**, 199–204.

Collins, J. C., Pilkington, T. C. and Schmidt-Nielsen, K. (1971) A model of respiratory heat transfer in a small mammal. *Biophys. J.* **11**, 886–914.

Crawford, E. C. (1962) Mechanical aspects of panting in dogs. *J. appl. Physiol.* **17**, 249–51.

Crawford, E. C. and Kampe, G. (1971) Resonant panting in pigeons. *Comp. Biochem. Physiol.* **40**A, 549–52.

Daniel, P. M., Dawes, J. D. K. and Prichard, M. M. L. (1953) Studies on the carotid rete and its associated arteries. *Phil. Trans. Roy. Soc. Lond. B*, **237**, 173–208.

Drorbaugh, J. E. (1960) Pulmonary function in different animals. *J. appl. Physiol.* **15**, 1069–72.

Fedde, M. R., Burger, R. E. and Kitchell, R. L. (1963) Electromyographic studies on certain respiratory muscles of the chicken. *Poult. Sci.* **42**, 1269.

Gifford, W. E. and McMahon, H. O. (1960) A new low-temperature gas expansion cycle. Part II. *Adv. in Cryog. Eng.* **5**, 368–72.

Gjönnes, B. and Schmidt-Nielsen, K. (1952) Respiratory characteristics of kangaroo rat blood. *J. cell. comp. Physiol.* **39**, 147–52.

Gregory, W. K. (1912) Notes on the principles of quadrupedal locomotion and on the mechanism of the limbs in hoofed animals. *Ann. N.Y. Acad. Sci.* **22**, 267–94.

Grigg, G. C. (1970) Water flow through the gills of Port Jackson sharks. *J. exp. Biol.* **52**, 565–8.

Hales, J. R. S. (1966). The partition of respiratory ventilation of the panting ox. *J. Physiol., Lond.* **188**, 45–6P.

Hales, J. R. S. and Bligh, J. (1969) Respiratory responses of the conscious dog to severe heat stress. *Experientia*, **25**, 818–19.

Hales, J. R. S. and Findlay, J. D. (1968) Respiration of the ox: normal values and the effects of exposure to hot environments. *Resp. Physiol.* **4**, 333–52.

Hall, F. G., Dill, D. B. and Guzman Barron, E. S. (1936) Comparative physiology in high altitudes. *J. cell. comp. Physiol.* **8**, 301–13.

Harwood, P. D. (1963) Therapeutic dosage in small and large mammals. Letter to the Editor, *Science, N.Y.* **139**, 684–5.

Hemmingsen, A. M. (1960) Energy metabolism as related to body size and respiratory surfaces, and its evolution. Copenhagen: *Reports of the Steno Memorial Hospital and Nordisk Insulinlaboratorium* **9**, 1–110.

Hinds, D. S. and Calder, W. A. (1971) Tracheal dead space in the respiration of birds. *Evolution*, **25**, 429–40.

Hutchens, J. O., Podolsky, B. and Morales, M. F. (1948) Studies on the kinetics and energetics of carbon and nitrogen metabolism of *Chilomonas paramecium*. *J. cell. comp. Physiol.* **32**, 117–41.

Jackson, D. C. and Schmidt-Nielsen, K. (1964) Countercurrent heat exchange in the respiratory passages. *Proc. natn. Acad. Sci. U.S.A.* **51**, 1192–7.

Jenkin, F. (1885) Gas- and Caloric-Engines. In *Heat and its Mechanical Applications*, pp. 101–74. London: Institution of Civil Engineers.

Kayser, Ch. and Heusner, A. (1964) Étude comparative du métabolisme énergétique dans la série animale. *J. Physiol., Paris* **56**, 489–524.

King, A. S. (1966) Structural and functional aspects of the avian lungs and air sacs. *Int. Rev. Gen. Exp. Zool.* **2**, 171–267.

King, A. S. and Cowie, A. F. (1969) The functional anatomy of the bronchial muscle of the bird. *J. Anat.* **105**, 323–36.

Kleiber, M. (1932) Body size and metabolism. *Hilgardia*, **6**, 315–53.

Kleiber, M. (1961) *The Fire of Life. An Introduction to Animal Energetics*. New York: John Wiley.

Kleiber, M. (1967) Prefatory chapter: An old professor of animal husbandry ruminates. *Ann. Rev. Physiol.* **29**, 1–20.

Krogh, A. (1929) *The Anatomy and Physiology of Capillaries*. 2nd edition. 422 pp. New Haven, Conn.: Yale University Press.

Kuhn, W., Ramel, A., Kuhn, H. J. and Marti, E. (1963) The filling mechanism of the swimbladder. Generation of high gas pressures through hairpin countercurrent multiplication. *Experientia* **19**, 497–511.

Landes, R. R., Leonhardt, K. O. and Duruman, N. (1964) A clinical study of the oxygen tension of the urine and renal structures. II. *J. Urol.* **92**, 171–8.

Larimer, J. L. and Schmidt-Nielsen, K. (1960) A comparison of blood carbonic anhydrase of various mammals. *Comp. Biochem. Physiol.* **1**, 19–23.

Lasiewski, R. C. and Calder, W. A. (1971) A preliminary allometric analysis of respiratory variables in resting birds. *Resp. Physiol.* **11**, 152–66.

Lasiewski, R. C. and Dawson, W. R. (1967) A re-examination of the relation between standard metabolic rate and body weight in birds. *Condor* **69**, 13–23.

Lasiewski, R. C. and Snyder, G. K. (1969) Responses to high temperature in nestling double-crested and pelagic cormorants. *Auk* **86**, 529–40.

Matyukhin, V. A. and Stolbow, A. J. (1970) Oxygen consumption in Baikal Lake omul (*Coregonus autumnalis migratorius*) and harius (*Thymallus arcticus P.*) in relation to swimming speed. In *Adaptatsiia Vodnykh Zhivotnykh Adaptatsiia k Usloviiam Gor i Gipoksii*. pp. 147–52. Novosibirsk, USSR: Akademiia Nauk SSSR, Sibirskoe Otdelenie.

Murrish, D. E. and Schmidt-Nielsen, K. (1970) Exhaled air temperature and water conservation in lizards. *Resp. Physiol.* **10**, 151–8.

Nayar, J. K. and Van Handel, E. (1971) The fuel for sustained mosquito flight. *J. Insect Physiol.* **17** 471–81.

Nielsen, B. and Nielsen, M. (1962) Body temperature during work at different environmental temperatures. *Acta physiol. scand.* **56**, 120–9.

Norris, K. S. (1953) The ecology of the desert iguana, *Dipsosaurus dorsalis*. *Ecology*, **34**, 265–87.

Odum, E. P., Rogers, D. T. and Hicks, D. L. (1964) Homeostasis of the nonfat components of migrating birds. *Science, N.Y.* **143**, 1037–9.

Prange, H. D. and Schmidt-Nielsen, K. (1970) The metabolic cost of swimming in ducks. *J. exp. Biol.* **53**, 763–77.

Raab, J. and Schmidt-Nielsen, K. (1971) Effect of activity on water balance of rodents. *Fedn. Proc.* **30**, 371.

Rayner, M. D. and Keenan, M. J. (1967) Role of red and white muscles in the swimming of the skipjack tuna. *Nature, Lond.* **214**, 392–3.

Riggs, A. (1960) The nature and significance of the Bohr effect in mammalian hemoglobins. *J. gen. Physiol.* **43**, 737–52.

Rikmenspoel, R., Sinton, S. and Janick, J. J. (˜969) Energy conversion in bull sperm flagella. *J. gen. Physiol.* **54**, 782–805

Rothschild, Lord (1953) The movements of spermatozoa. In *Mammalian Germ Cells.* pp. 122–33. Ciba Foundation Symposium. London: J. & A. Churchill Ltd.

Rubner, M. (1883) Ueber den Einfluss der Körpergrösse auf Stoff-und Kraftwechsel. *Z. Biol.* **19**, 535–62.

Salt, G. W. (1964) Respiratory evaporation in birds. *Biol. Rev.* **39**, 113–36.

Scheid, P. and Piiper J. (1970) Analysis of gas exchange in the avian lung: theory and experiments in the domestic fowl. *Resp. Physiol.* **9**, 246–62.

Schmidt-Nielsen, K. (1964) *Desert Animals. Physiological Problems of Heat and Water.* 277 pp. London: Oxford University Press.

Schmidt-Nielsen, K. (1970) *Animal Physiology*, 3rd edition. 145 pp. Englewood Cliffs, N.J.: Prentice-Hall, Inc.

Schmidt-Nielsen, K., Bretz, W. L. and Taylor, C. R. (1970a) Panting in dogs: unidirectional air flow over evaporative surfaces. *Science, N.Y.* **169**, 1102–4.

Schmidt-Nielsen, K., Hainsworth, F. R. and Murrish, D. E. (1970b) Countercurrent heat exchange in the respiratory passages: effect on water and heat balance. *Resp. Physiol.* **9**, 263–76.

Schmidt-Nielsen, K., Kanwisher, J., Lasiewski, R. C., Cohn, J. E. and Bretz, W. L. (1969) Temperature regulation and respiration in the ostrich. *Condor* **71**, 341–52.

Schmidt-Nielsen, K. and Larimer, J. L. (1958) Oxygen dissociation curves of mammalian blood in relation to body size. *Am. J. Physiol.* **195**, 424–8.

Schmidt-Nielsen, K. and Pennycuik, P. (1961) Capillary density in mammals in relation to body size and oxygen consumption. *Am. J. Physiol.* **200**, 746–50.

Schmidt-Nielsen, K., Schmidt-Nielsen, B., Jarnum, S. A. and Houpt, T. R. (1957) Body temperature of the camel and its relation to water economy. *Am. J. Physiol.* **188**, 103–12.

Scholander, P. F. (1958) Counter current exchange. A principle in biology. *Hvalrådets Skrifter*, No. 44, 1–24.

Scholander, P. F. and Van Dam, L. (1954) Secretion of gases against high pressures in the swimbladder of deep-sea fishes. 1. Oxygen dissociation in blood. *Biol. Bull.* **107**, 247–59.

Scholander, P. F. & Krog, J. (1957) Countercurrent heat exchange and vascular bundles in sloths. *J. appl. Physiol.* **10**, 405–11.

Soum, J. M. (1896) Récherches physiologiques sur l'appareil réspiratoire des oiseaux. *Ann. Univ. Lyon*, **28**, 1–126.

Stahl, W. R. (1972) *Physiological Similarity and Modeling. The Application of Dimensional Analysis and Physical Similarity Theory to Mammalian Physiology.* New York: Appleton-Century-Crofts. In the press.

Steno, N. (1664) *De musculi et glandulis.* Amstelodami. (Quoted from Broman, I. (1921) *Z. ges. Anat.* **60**, 439.)

Suckling, J. A., Suckling, E. E. and Walker, A. (1969) Suggested function of the vascular bundles in the limbs of *Perodicticus potto. Nature, Lond.* **221**, 379–80.

Swift, J. (1726) *Travels into Several Remote Nations of the World. In Four Parts. By Lemuel Gulliver. First a Surgeon, and then a Captain of Several Ships.* London: Benj. Motte.

Taylor, C. R. (1969) The eland and the oryx. *Sci. Am.* **220**, 88–95.

Taylor, C. R. (1970) Dehydration and heat: effects on temperature regulation of East African ungulates. *Am. J. Physiol.* **219**, 1136–9. ..

Taylor, C. R., Dmi'el, R., Fedak, M. and Schmidt-Nielsen, K. (1971a) Energetic cost of running and heat balance in a large bird, the rhea. *Am. J. Physiol.* **221**, 597–601.

Taylor, C. R., Schmidt-Nielsen, K., Dmi'el, R. and Fedak, M. (1971b) Effect of hyperthermia on heat balance during running in the African Hunting Dog. *Am. J. Physiol.* **220**, 823–7.

Taylor, C. R., Schmidt-Nielsen, K. and Raab, J. L. (1970) Scaling of the energetic cost of running to body size in mammals. *Am. J. Physiol.* **219**, 1104–7.

Tennent, D. M. (1946) A study of water losses through the skin in the rat. *Am. J. Physiol.* **145**, 436–40.

Tenney, S. M. and Remmers, J. E. (1963) Comparative quantitative morphology of the mammalian lung: diffusing area. *Nature, Lond.* **197**, 54–6.

Thomas, S. P. and Suthers, R. A. (1970) Oxygen consumption and physiological responses during flight in an echolocating bat. *Fedn. Proc.* **29**, 265.

Tucker, V. A. (1968a) Respiratory physiology of house sparrows in relation to high-altitude flight. *J. exp. Biol.* **48**, 55–6.

Tucker, V. A. (1968b) Respiratory exchange and evaporative water loss in the flying budgerigar. *J. exp. Biol.* **48**, 67–87.

Tucker, V. A. (1970) Energetic cost of locomotion in animals. *Comp. Biochem. Physiol.* **34**, 841–6.

Tucker, V. A. (1971) Flight energetics in birds. *Am. Zool.* **11**, 115–24.

Verzár, F., Keith, J. and Parchet, V. (1953) Temperatur und Feuchtigkeit der Luft in den Atemwegen. *Pflüger's Arch. ges. Physiol.* **257**, 400–16.

Von Hoesslin, H. (1888) Ueber die Ursache der scheinbaren Abhängigkeit des Umsatzes von der Grösse der Körperoberfläche. *Du Bois-Reymond Arch. Anat. Physiol.* 323–79.

Walker, J. E. C., Wells, R. E. and Merrill, E. W. (1961) Heat and water exchange in the respiratory tract. *Am. J. Med.* **30**, 259–67.

Webb, P. (1951) Air temperatures in respiratory tracts of resting subjects in cold. *J. appl. Physiol.* **4**, 378–82.

Webb, P. (1955) Respiratory heat loss in cold. *Fedn. Proc.* **14**, 486.

Weis-Fogh, T. (1952) Weight economy of flying insects. *Trans. Ninth Int. Congr. Ent.* **1**, 341–7.

Wohlschlag, D. E., Cameron, J. N. and Cech, J. J. (1968) Seasonal changes in the respiratory metabolism of the pinfish (*Lagodon rhomboides*). *Contr. Marine Sci.* **13**, 89–104.

Zeuthen, E. (1942) The ventilation of the respiratory tract in birds. *Kl. danske Vidensk. Selsk. Biol. Medd.* **17**, 1–50.

Index